The Honey Bee Solution to Varroa

A practical guide for beekeepers

Steve Riley

Chair and Education Officer

Westerham Beekeepers

Northern Bee Books

The Honey Bee Solution to Varroa
Copyright © Steve Riley

All rights reserved. No part of this publication may be reproduced, stored in a retrieval system, transmitted in any form or by any means electronic, mechanical, including photocopying, recording or otherwise without prior consent of the copyright holders.

Published 2024 by
Northern Bee Books,
Scout Bottom Farm,
Mytholmroyd,
West Yorkshire
HX7 5JS (UK)
Tel: 01422 882751
Fax: 01422 886157
www.northernbeebooks.co.uk

ISBN 978-1-914934-77-3

Design and artwork DM Design and Print

Cover:
© Dan Peterson's illustration frames the debate between natural selection and creationism in the nineteenth century. www.danpetersonillustrator.co.uk

Down House, Darwin's family home, is situated in the northern tip of Westerham Beekeepers' membership area.

The Honey Bee Solution to Varroa

A practical guide for beekeepers

Steve Riley
Chair and Education Officer
Westerham Beekeepers

Foreword

The *Varroa destructor* ("Varroa") mite has been the major pest of honey bees over the past 30 years in the UK. After Varroa arrived, beekeepers initially were very reluctant to use miticides to control the mite population. However, now putting chemicals into honey bee colonies to control *Varroa destructor* is just an accepted part of beekeeping, but is this the only option?

During the past 10-15 years there has been a quiet, mostly hidden, revolution in UK beekeeping. This involves increasing numbers of beekeepers managing their colonies without the need for any type of Varroa treatment. In the UK, it was the late Ron Hoskins that led the way and maintained his colonies for decades without treating for Varroa. Currently there are an estimated 1,800 UK beekeepers that haven't treated for Varroa for six years or more. So, what is going on and how can you save money and time by becoming one of the growing number of beekeepers that don't need to treat for Varroa?

This book shows you the way. It's written by a beekeeper for beekeepers. It's based on the author's personal journey of becoming a treatment free beekeeper with respect to Varroa. The first few chapters introduce the reader to the basic biology of Varroa, the role of viruses and the emergence of Varroa-resistance both in the UK and further afield. This information allows the reader to fully understand how resistant-honey bee populations have adapted to deal with these mites. Much of this information was not known when the author's selection program started since research in Varroa-resistance has developed rapidly during the past six years. Thus, the first few chapters bring the reader bang up-to-date with the latest science. This provides the foundation that allows the beekeeper to understand the 'why' that lies behind the more practical and important second part of the book. This deals with how the author and his colleagues successfully selected, over a six-year period, for a Varroa-resistant honey bee population, using simple methods available to all beekeepers.

Their journey was not a straight-forward one; they made mistakes and learning curves were steep. Experience is learning from mistakes and chapter 14 'Lessons learned' contains lots of key insights. This is a very practical book which includes the author's observations and ideas, whilst addressing many of the concerns beekeepers have voiced about transitioning to treatment free beekeeping. The book is richly illustrated and importantly is transparent in that it provides their data and observations on which decisions were made. This is the first book that provides a practical long-term treatment free solution to the Varroa problem. The methods used are simple, easy to follow and adaptable to each beekeeper or associations' own situation and it works with any race of honey bee, kept in any hive type, in any type of environment.

Best of luck on your personal journey to managing your own Varroa-resistant honey bees.

Emeritus Professor Stephen Martin F.E.S.

Contents

1. Introduction — 1
2. Emergence of Varroa-resistance — 3
3. Varroa and viruses; why they overwhelm colonies — 7
4. How Varroa-resistant colonies manage their mite populations — 15
5. Why selecting for Varroa-resistant traits is important — 30
6. Westerham Beekeepers' monitoring of Varroa-resistance traits — 33
7. Other approaches to monitoring — 48
8. Seasonality of honey bees' hygienic behaviour against Varroa — 55
9. Contributory colony health factors — 72
10. Selection process and spreading Varroa-resistance — 78
11. Drones: a pivot for Varroa-resistant traits — 86
12. Sustainable apiary for Varroa-resistant bees — 90
13. Honey yields and colony survival — 97
14. Lessons learned — 102
15. Getting started / transitioning from treated colonies — 109

Acknowledgements — 117

References and further reading — 119

Appendix — 126

Short-term biotechnical stepping-stone — 126

1
Introduction

A group of us at Westerham Beekeepers, a club in the Kent / Surrey borders of England, became interested in Varroa-resistant bees when it became clear that *Apis mellifera*, the European honey bee, was able to survive the presence of *Varroa destructor* ("Varroa") without intervention from the beekeeper. Like many beekeepers, we had always been uncomfortable applying pesticide controls for Varroa mites in our colonies and looked for and found a long-term, bee-led solution.

When we started the project in 2017, there was already a growing number of well-established naturally Varroa-resistant colonies in the UK and elsewhere. The challenge for us was the lack of an educational pathway to follow and a full understanding of the bees' mechanisms of how they were controlling their mite populations. The bees were ahead of the scientists researching them, who were in turn, a country mile ahead of beekeeper education and training. We decided to embark on collecting our own data and research to inform selection decisions affecting our colonies.

Currently, most beekeeper education focuses on how to reduce Varroa numbers to allow colonies to survive. A strategy, known as Integrated Pest Management, was a sensible and essential approach to maintaining honey bee colonies in the early days of Varroa arriving. Over 30 years on, these Varroa reduction methods have the unintended consequence of perpetuating the Varroa problem.

Understanding why honey bees are surviving Varroa has attracted the attention of the world's finest scientists in apiculture. We were fortunate to have Dr. Ralph Büchler as our mentor for the start-up of the project and in recent years, Emeritus Professor Stephen Martin. Thank you both.

This book combines the steps we took to identify Varroa-resistant traits with the latest research underpinning the bees' mechanism for controlling mite populations. It is intended to bridge the gap in education between the research on the survival mechanisms that honey bees are deploying against Varroa, and the practical steps beekeepers can take to identify these observable traits. Also included are the lessons we learned on the way, a whole year's analysis of a resistant colony, and a step-by-step approach to getting started and transitioning off miticide treatments. The book finishes with evidence of bees which have restricted Varroa reproduction in worker brood to such a low level, that it is akin to how *Apis cerana* in Asia, the original host, manages the parasite.

At Westerham Beekeepers, using our local honey bees, we looked for and found colonies that manage their own Varroa populations without the need for beekeeper intervention. The methods are taught at our Training Apiary and shared with other interested clubs who have started their own projects. Finding Varroa-resistance is achievable through monitoring small numbers of locally adapted colonies of honey bees.

We hope you find the book useful. Selecting traits for Varroa-resistance heralds the biggest area of bee improvement for many decades and is within the grasp of all beekeepers.

Steve Riley

Westerham Beekeepers

2
Emergence of Varroa-resistance

Since the 1990's *Apis mellifera* honey bees have been adapting to Varroa in many countries, particularly those where mite treatment was unavailable or too costly. Large, productive and healthy Varroa-resistant honey bee populations exist throughout sub–Saharan Africa, most of South & Central America and Cuba. There are well studied populations in Avignon in France, Gotland Island in Sweden, Oslo in Norway and the Arnot Forest in New York State, which are providing valuable insights into the traits the bees have developed to combat the mite. A good (free to access) summary of the history of resistant colonies was published by Barbara Locke in 2016: "Natural *Varroa* mite – surviving *Apis mellifera* honey bee populations."

In short, Varroa-resistance has occurred where beekeepers have not interfered with the process of adaption.

Varroa in the UK

Varroa mites were first reported in the UK in 1992, having spread gradually across continents from Asia. In older publications, the mites were referred to as *Varroa jacobsoni* but are now recognised as *Varroa destructor* (Anderson and Trueman 2000). UK honey bees originally had none of the defences against Varroa that *Apis cerana*, the Asian honey bee, had developed to allow the parasite and host to exist in equilibrium.

During the first five years, colony losses were high across the UK, leading to many beekeepers giving up. Beekeepers were initially very reluctant to put chemical miticides in their colonies and tried various weird and wonderful methods of mite control. The entire situation was made worse by the large-scale collapse of unmanaged / feral colonies causing large numbers of Varroa mites to invade nearby colonies. At that time, the role of viruses in the collapse of the colonies was unknown and caused initial confusion among beekeepers, since some colonies appeared healthy despite heavy mite infestations, while

2 Emergence of Varroa-resistance

others collapsed despite much lower infestation levels.

After the initial period of high colony losses, beekeepers started to use the highly efficient miticides Apistan® (tau-fluvalinate) and Bayvarol® (flumethrin) to control mite populations. Varroa became resistant to these compounds, which was first discovered in 2001 and later became widespread. This caused beekeepers to switch to other products based on chemicals such as thymol, formic and oxalic acid. Most beekeepers currently treat once or twice a year, occasionally multiple times, with little or minimal monitoring of mite levels.

However, much has changed in the 30+ years since Varroa's arrival in the UK. Unmanaged / feral populations have rebounded (Visick and Ratnieks 2023), mite reinvasion rates are lower and we are seeing the emergence of colony survival rates on long-standing, non-treated colonies matching those of treated bees.

The UK pioneer of selecting for resistant colonies was the late Ron Hoskins, founder of the Swindon Honey Bee Conservation Group. He ceased the use of miticide treatments in 1995 and after losing many of his original colonies, commenced a selection program using those that remained, which after years of honing, focused on signs of:-

- Damaged mites on the bottom board, surmising they had been bitten by bees and groomed off.
- Uncapping and ejecting of damaged pupae.

In all other respects, Ron's beekeeping was conventional. He used British National hives, foundation, intervened with queens, selected for temperament and took a honey crop. He was a beekeeper at the forefront of understanding resistant traits in Varroa and well ahead of his time.

The largest known area of Varroa-resistant bees in the UK can be found in north-west Wales. Around 100 Lleyn & Eifionydd BKA beekeepers have some 500 Varroa-resistant colonies spread across a 2500km^2 region where treatments ended from 2009 onwards (Hudson and Hudson 2020). They noticed the return of long-lived colonies in trees and cavities in buildings from which they collected swarms and subsequently stopped all treatments. The wild/feral colonies are the likely genetic source of Varroa-resistance in the area.

Fig. 2.1: Distribution map of treatment-free honey bee colonies in north-west Wales, March 2019. The largest area of Varroa-resistance in the UK.

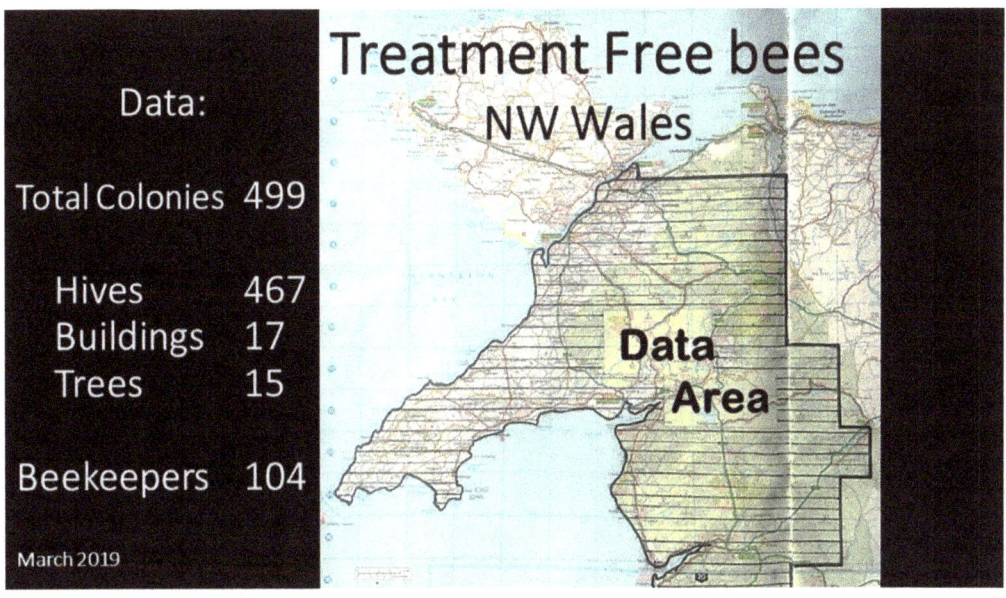

Map image © AA Media Ltd 2009 All Rights reserved.

In addition to these Welsh colonies, Varroa-resistant populations have appeared throughout the UK. A survey of UK beekeepers' Varroa treatment habits (Valentine and Martin 2023) included responses from 242 beekeeping associations. From this an estimated 1,800 beekeepers had not used miticide treatments for at least six years with a broad geographic spread of non-treating beekeepers around England and Wales, although curiously not Scotland.

Varroa-resistance occurred quicker than many predicted

There is an oft repeated myth in beekeeping circles that adaption, or the evolution of resistance to Varroa is on a timeframe outside of our lifetimes. This is a common misperception. The pace at which Varroa-resistance occurs in honey bees depends on the level of pressure from natural selection. For example, in South Africa and Cuba it took around 5-8 years for Varroa-resistance to be widespread after the decision was taken not to treat after Varroa arrived, but there were up-front losses (Allsopp 2006, Luis *et.al* 2022). This is why there is great interest in securing bees from long-standing unmanaged colonies from cavities in the UK and elsewhere. Their exposure to natural selection encourages Varroa-resistant traits to be developed in the local population and spread around the area.

Honey bees already have similar hygienic behaviours for other brood disorders and pests, including the uncapping of worker brood for the presence of wax moth, and uncapping and removal where there is chalkbrood and sacbrood virus. For a deeper insight into the subject, I would recommend: "Perspectives on hygienic behaviour in *Apis mellifera* and other social insects", Spivak and Danka 2021.

3
Varroa and viruses
Why they overwhelm colonies

First, a little Varroa biology which will help with the understanding of later chapters on the mechanisms honey bees are deploying against Varroa.

Varroa reproduction (in colonies with low defences to Varroa)

The mite's egg laying cycle commences after the foundress mite enters an open brood cell up to 48 hours before it is sealed. Once the cell is sealed, she creates a feeding hole on the larvae for herself and future offspring.

The first egg is laid about 60 hours after the cell is sealed. This always develops into a son whose only role is to mate with his sisters. He looks a lot like a younger version of his sisters (see Fig. 3.1).

Subsequent eggs are then laid at around 30-hour intervals, which are the daughter mites. Once sexually mature, the son fertilises his sisters. The timeframe for successful mating is tight, especially in worker cells where the sealed pupation period is 12 days. On average about 1.8 (Donze *et al.* 1996) viable daughters are produced in worker brood cells. In drone brood, due to the longer pupation period (14 versus 12 days), about 3 viable daughters are produced each cycle (Martin 1998). In both cases, the mother mite survives and exits the cell with her daughters. These are the outcomes in mite-susceptible colonies, i.e. those with low mite defences. Resistant colonies interrupt this process.

3 Varroa and viruses

Fig. 3.1: A Varroa mite family.

Upper row from left to right: adult foundress and daughters; deutochrysalis (and moult), deutonymph, protonymph.

Lower row from left to right: adult male, deutonymph, protonymph.

The brick-brown foundress mite can be almost 2mm wide and the whole mite family can be seen on Varroa insert boards, with the help of magnification.

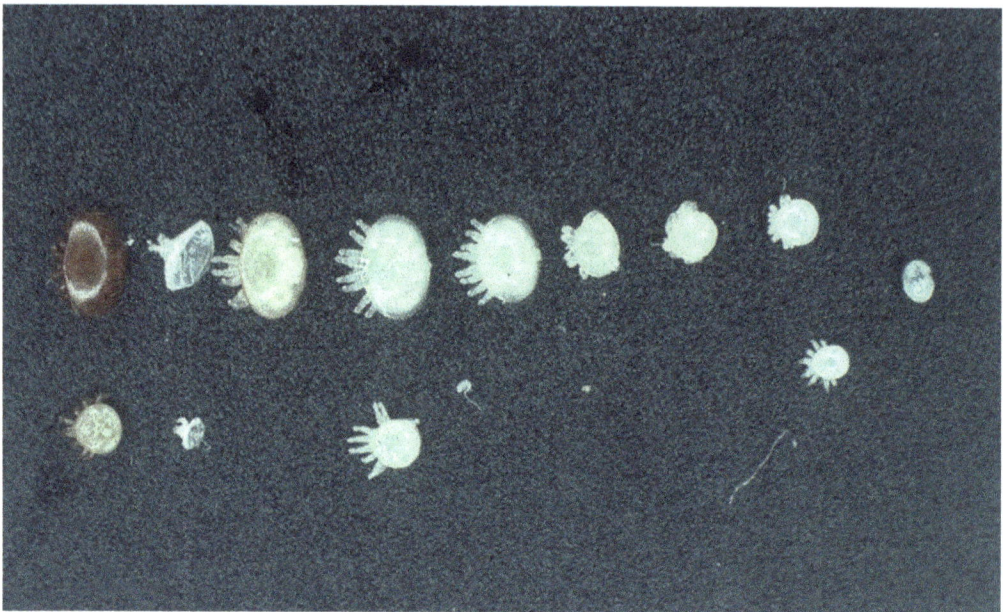

Picture: Prof S Martin.

In *Apis mellifera*, Varroa can reproduce in both worker and drone brood:

In *Apis cerana* (the Asian honey bee), Varroa almost exclusively reproduces in drone brood since bees remove any mite-infested worker brood. Drone brood is seasonal for *Apis cerana*, which limits the opportunity for Varroa to reproduce. In addition, multiple mite-infested drone cells have a high mite mortality rate and the result is a natural balance in the colony between host and parasite.

However, when Varroa came to the UK, our honey bees (*Apis mellifera*) had little or no defences to the mites breeding in both the drone and the more populous worker brood.

In the Westerham Beekeeper colonies, worker brood begins to be laid after the winter brood break in December for the season ahead. This is the first breeding opportunity for overwintering Varroa and if uninterrupted, foundress mites will breed in worker brood (and in seasonal drone brood) for the next 11 months.

Fig. 3.2:   Illustrates the levels of worker versus seasonal drone brood in one of the Westerham Beekeepers' colonies.   Reproduction opportunities for Varroa are overwhelmingly in worker brood and this is the key battle ground.

The dip in brood reflects the brood-break after a reproductive swarm.

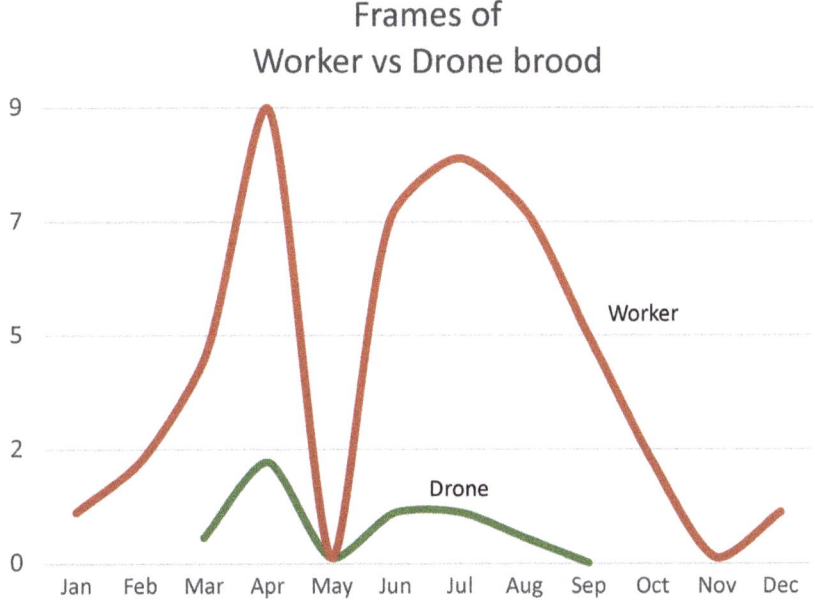

Chart: Steve Riley.

A foundress (mother) mite has a finite reproductive capacity:

Her egg laying capacity is estimated to range from 18-24 eggs but as high as 30 – this will depend on how well she was mated. There are 2-3 reproductive cycles (Rozenkranz *et al.* 2010); during each cycle, the foundress lays four or five eggs in worker sealed brood and five or six eggs in drone sealed brood. However, in *Apis mellifera* worker sealed brood many of the eggs fail to produce mature daughters since Varroa has not evolved to reproduce in the shorter pupation period of worker brood, i.e. 12 days versus 14 days in the *Apis cerana* drone brood.

The foundress mite can live up to one year:

Varroa's natural breeding ground, the drone brood of *Apis cerana* in Asia, is only present for around two months a year. When not reproducing, the mothers are in the dispersal phase (previously referred to as phoretic) living and feeding (Ramsey *et al.* 2019) on the adult honey bees. This explains how foundress mites can overwinter on our bees.

Viruses, honey bees and Varroa

Viruses can be 'head-swivelling' complex and an inaccessible area for most of us beekeepers. They are widespread in our colonies, more than is generally appreciated, as shown by the read-through from research across Belgium.

Fig 3.3: Research across 557 colonies in Belgium found that all of the colonies had deformed wing virus, black queen cell virus and sacbrood virus (Matthijs *et al.* 2020):

Virus	% positive colonies in Flanders	% positive colonies in Wallonia
Acute Bee Paralysis virus	8.8	26.5
Black Queen Cell virus	99.23	99.65
Chronic Bee Paralysis virus	29.23	30.27
Deformed wing virus – A	18.07	37.75
Deformed wing virus – B	100	100
Sacbrood virus	97.67	96.26

Mostly, these viruses exist at low levels and are handled by the bees' immune defence barriers.

What we are concerned about is virulence (i.e. the severity of the disease symptom in the bees), rather than virus existence. In virologists' jargon, "load" is much more important than "prevalence", particularly when it comes to deformed wing virus.

Deformed wing virus ("DWV")

DWV was first identified in 1982 by Leslie Bailey and Brenda Ball at the Rothamsted Research Centre in the United Kingdom (Martin and Brettell 2019), but was rarely noticed by beekeepers or scientists until several years after Varroa arrived.

Naturally, without Varroa, DWV can be transmitted from bee to bee via a series of mechanisms:-

- trophallaxis (bees feeding each other)
- nurse bees feeding larvae
- cleaning out DWV infected larvae and pupae
- from queen to eggs
- via drone sperm to virgin queens
- pollen

Through these many transmission routes, which have existed for a long time, the virus would be present in a colony at very low viral loads as a symptomless infection. DWV was dealt with by the bees' immune system and beekeepers knew of, or cared little about it. It still remains at low, asymptomatic levels where there is no presence of Varroa, e.g. Isle of Colonsay, Scotland (Evans 2019).

The arrival of Varroa into *Apis mellifera* colonies elevated both the prevalence and load of DWV to threatening, pathogenic levels.

Researcher Nadine Möckel discovered the key. She found that <u>feeding</u> modest amounts of DWV to pupa or adult bees resulted in covert infections (i.e. not noticeable to a beekeeper). However, <u>injecting</u> small amounts of the virus caused overt infections (Möckel *et al*. 2011). Replace injecting virus by Varroa when they feed on larvae and pupae in sealed brood and this is where small particles of virus rapidly increase within the developing bee (Dainat 2012).

The key impact of Varroa transmitted DWV is on the bees' longevity. Sealed brood that are injected with DWV by feeding Varroa have their life span reduced by about two thirds. If Varroa feeding occurs <u>only</u> on the adult bee, there is a slight reduction in longevity (Martin and Grindrod 2020).

There are obvious implications here for colony survival over winter in the UK and other northern hemisphere countries, where winter bees must live 5-6 months for the colony to endure and thrive in the spring.

Fig. 3.4: Deformed wing virus is endemic in colonies but only rarely expressed in wing deformation. Lifespan, learning and memory are still affected.

Picture: Courtesy of The Animal and Plant Health Agency (APHA), Crown Copyright.

There have been four types of DWV recorded. Type-C is rare and type-D no longer exists. Type-A was prevalent in the UK but type-B has taken over and now dominates (Kevill *et al*. 2019). What does this mean for beekeepers?

Emeritus Professor Stephen Martin sums it up:

"DWV-B is more transmittable than DWV-A. This is because unlike DWV-A, DWV-B can replicate within the Varroa mite. Both strains can kill a colony, with some evidence suggesting DWV-B is more virulent, but more studies need to be conducted." (www.varroaresistant.uk)

How Varroa overwhelm mite-susceptible colonies

This is a subject where, unusually in beekeeping, there is a consensus. The short answer is that high Varroa numbers elevate DWV to clinically pathogenic levels. The infestation of Varroa in worker brood peaks in the late summer just at the time when the long-lived winter bees are being raised. If enough worker bees have been infected with DWV via

3 Varroa and viruses

the mite during their pupa development, this will lead to the colony's demise during the winter.

Fig. 3.5: Shows the development of brood and mite infestation in a (non-swarming) colony with <u>low defences</u> against mites. Critical risk occurs in the autumn where there are too many mites on the winter bees.

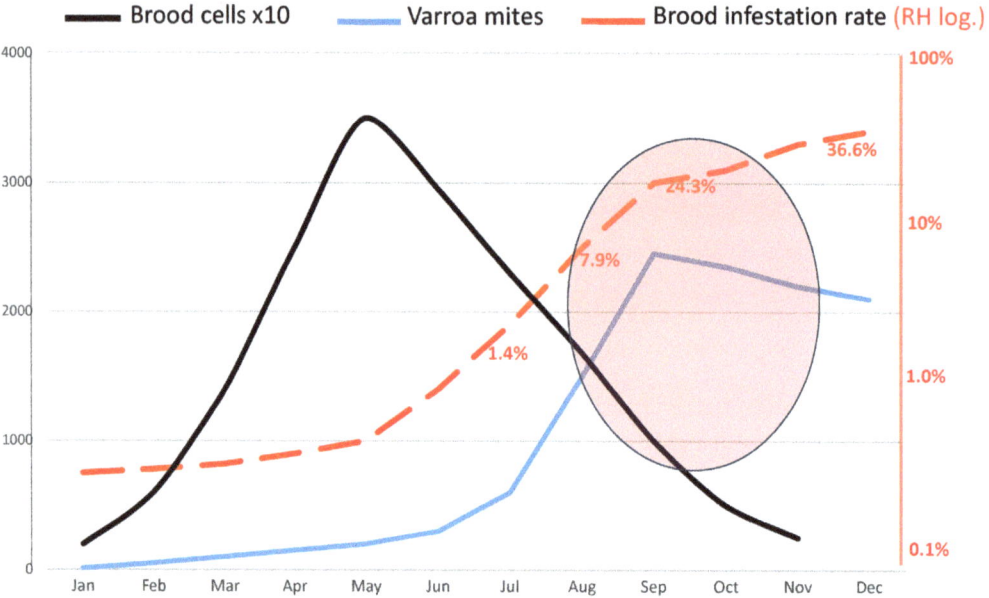

Re-drawn graph with credit to, and permission from Dr. Ralph Büchler

The colony's demise can be a surprise to the beekeeper. During the last autumn inspections, the hive looked healthy and was full of summer bees and recently laid winter bees; the latter will have high levels of DWV, which is often not appreciated as it is rarely expressed in the wings. The summer bees naturally die off in the ensuing months. The winter bees, with their DWV reduced longevity, are dead by around the end of the year, leaving a diminished colony unable to thermoregulate with the cold winter weather.

Our interest is in: *How naturally resistant honey bees survive this fate*. The mechanisms for doing so are discussed in the next chapter.

4
How Varroa-resistant colonies manage their mite populations

Resistance is the ability not to be affected by something, especially adversely.

Part of the beekeeper's problem has been an unclear definition of Varroa-resistance which is often used interchangeably with Varroa-tolerance. It is a nebulous phrase that sounds highly desirable but out of reach. We needed to understand the mechanisms of how it worked.

The darkness of the hives holds many secrets of the honey bee from the beekeeper, so we are indebted to scientists for unearthing the bees' hygienic behaviour against Varroa from research on long-term naturally resistant colonies. The good news is that the complex science is handled by the bees and the resistant traits are mostly, easy to observe for beekeepers.

The key to Varroa-resistance is the bees' ability to limit growth in the Varroa population through the season. We also include Virus tolerance in our wider definition, which is discussed at the end this chapter. We used to believe that grooming and mite mutilation were important, but after much colony research, came to a painful conclusion that there wasn't a strong correlation with bees' control of Varroa – more on this in Chapter 14, Lessons learned!

Bees limit the growth in Varroa numbers and therefore, the vectors of virus, by interrupting Varroa reproduction through the detection and removal of mite-infested worker brood, which falls into 3 phases:-

1) The search for Varroa families and parasitised pupae
2) Removal of parasitised pupae preventing Varroa reproduction
3) Recapping brood cells with healthy pupae

These are discussed in more detail below.

1) The search for Varroa families and parasitised pupae

Patrolling nurse bees check temperature, humidity and respiratory conditions by touching the micro porous surface of brood cells with their antennae. If you take a video of the brood area in your colonies, you can see this occurring. There are thousands of sensilla (nerve receptors) on each antenna, which also includes an olfactory (smell) ability capable of differentiating healthy from unhealthy brood. Researchers have already shown that honey bees can detect spores of American foulbrood under cappings (Rothenbuhler 1964) and as beekeepers, we see cells uncapped for chalkbrood (Gilliam *et al.* 1983), sacbrood virus and burrowing wax moth larva (Villegas and Villa 2006). Identifying issues under a cell capping is not a new hygienic behaviour for honey bees, but it has been adapted for Varroa detection.

Work from Professor Marla Spivak's group (Spivak *et al.* 2003) uncovered the role of different workers in the uncapping process and changes in brain chemistry of specialised detector worker bees. Thus, the honey bees already had a detection mechanism, but just needed time to learn to associate the unique smell coming from mite-infested cells with the presence of Varroa.

The first stage of the uncapping process is for the bees to create a pin-prick size hole in the cell cap, which allows further investigation. Other nurse bees are involved in opening and enlarging the cell, and then removing the pupa. From monitoring an observation hive, researchers recorded the average age of the detector, uncapping and chewing-out bees to be 11.13 ±3.7 days, with a range of 6 to 18 days old (Mondet *et al.* 2015). Further research is required to confirm these behaviours as developmental with age, i.e. temporal polyethism.

The research of Fanny Mondet and team (Mondet *et al.* 2021) identified 6 chemical cues, unique to brood cells that have been infested by Varroa, that trigger a hygienic response; in this case, uncapping the cell to investigate further. These compounds are diffused through the cell capping and identified in the bees' antennal lobes. Resistant bees have learned to recognise odours of Varroa and parasitised brood, whereas Varroa naïve or susceptible bees cannot do so to the same level.

Fig. 4.1: Research identified chemical cues leading to uncapping.

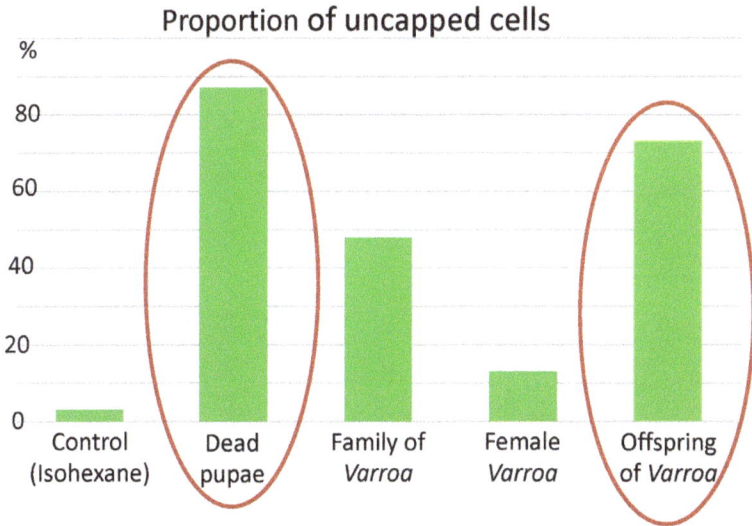

Graph: Redrawn with credit to Mondet *et al.* 2021: Chemical detection triggers honey bee defense against a destructive parasitic threat.

It is perhaps not surprising that dead or parasitised pupae lead to the uncapping of over 80% of cells but worth noting that Varroa rarely kill the pupae. The really interesting discovery was that bees could detect Varroa offspring inside a brood cell, leading to over 70% of cells being uncapped. Also of note, was the lack of uncapping of a single (non-reproductive) female mite, assumed to be infertile. This is a common feature of Varroa-resistant colonies.

Beekeepers can easily observe the fully opened cell, with a small lip around the edge, assumed to be capping held in reserve for reuse.

4 How Varroa-resistant colonies manage their mite populations

Fig. 4.2: Uncapping at the pink to purple-eyed stage (often occurs in groups).

Picture: Steve Riley.

The picture above is typical of a cluster of uncapped cells seen in Varroa-resistant colonies, in this case, one from Westerham Beekeepers. It is a grouped cluster as opposed to a straight line, which tends to be wax moth. In the cluster group, there will be a small number of cells with Varroa; but not all of them will be infested. The chemical odours emanating from the cell caps are wafted around by bee movements, leading to a level of inaccuracy in the bees' targeting of which cells to uncap (Grindrod and Martin 2021), resulting in a cluster pattern.

Fig. 4.3: Bees uncapping for wax moth larvae leave a straight line as opposed to clusters of uncapped cells when seeking Varroa.

Picture: Courtesy of The Animal and Plant Health Agency (APHA), Crown Copyright.

2) Removal of parasitised pupae preventing Varroa reproduction

Brood removal of an infested pupae occurs at the pink to purple-eyed stage. You can see in Fig. 4.4 bees chewing-out a pupa in one of the author's colonies – sometimes you are fortunate enough to have a camera phone handy. During the video, from which this still was taken, up to four different bees were seen chewing-out the pupa.

4 How Varroa-resistant colonies manage their mite populations

Fig. 4.4: Bees collectively chewing-out a pupa.

Picture: Steve Riley.

It's worth noting that hygienic colonies with this trait of removing parasitised pupae will have pop holes in the brood, visible in Fig. 4.4. We like to see this and consider it normal for a Varroa-resistant brood pattern. Frames of wall-to-wall worker brood with virtually no misses are likely to result from bees with no or low defences to mites – we view those colonies as Varroa breeders!

Whilst the chewing-out of infected pupae can transfer deformed wing virus, this is dealt with through the normal immune system of the bee, as opposed to injecting the virus at the larval or pupa stage, where the damage results. Ejecting a pupae ladened with DWV from a colony reduces the virus load (a little), with the main benefit being the reduction in future mites to vector the DWV.

Timing of the bees' interruption is key

We need to look at the reproduction stages of Varroa in worker brood cells to see why removing the pupa interrupts mite reproduction. As discussed earlier, most opportunities for Varroa occur in worker brood through the season – that's the key battleground.

Fig. 4.5 Illustrates the timing of bees' intervention to stop mite reproduction.

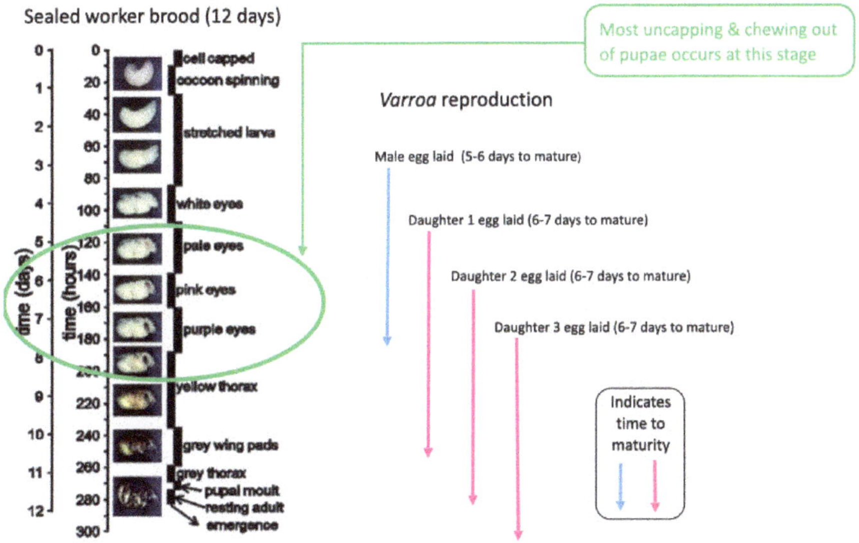

Chart: Steve Riley. Picture: S Camazine (Human *et.al* 2013).

In Fig. 4.5, the foundress mite has already entered the cell and hides in the larval food ahead of the cell being capped. Post-capping, she creates a feed hole by piercing the cuticle of the larvae for herself and her family. Her oogenesis (egg creation) cycle is triggered either by the chemical cues from the larva or the mite feeding process.

After about 60-70 hours, the first egg is laid, destined to be a male, followed at c.30-hour intervals by 4-5 female eggs. These eggs develop through the juvenile stages (larva, protonymph and deutonymph) and reach maturity after 5-6 days for the son and 6-7 days for the daughters (Donze and Guerin 1994: Rehm and Ritter 1989).

Once sexually mature, the son will mate with his sexually mature sisters. This commences around the 10th day of the 12 days that worker brood is sealed, so very much at the end of the pupation process and just ahead of the adult bee emerging.

Mating is the male's only role, and he dies together with any daughter mites who fail to reach maturity.

On the 12th day when the adult bee emerges, around 1.8 successfully mated daughters on average, plus the mother mite, leave the cell. This is the expected outcome for bees that have **no or low defences against Varroa**, resulting in substantial growth in mite numbers through the season.

Resistant bees have learnt to detect Varroa offspring / parasitised pupae and by removing them, interrupt the reproduction of the mites. Uncapping the cell and chewing-out infected pupa occurs at the pinkish to purple-eyed stage of the pupa's development, which is before the Varroa son and daughters are sexually mature enough to mate (see Fig. 4.5). **Bees interrupting the reproduction of mites is the key to Varroa-resistance**. It stops both the current mite reproduction from taking place, but also the future reproduction, through the creation of a cohort of poorly mated or infertile female mites.

Infertile mites:

A consequence of the bees' interference in the Varroa reproduction cycle is a substantial increase in the number of infertile foundress mites in the colony. This is due to two primary factors:-

1. Foundress mites exhausting their egg laying capacity through the interrupted reproductions, although they can escape when bees chew-out the pupae. Mother mites only reproduce between 2 and 3 times in the 2-3 months reproductive phase of their lives when there is brood in the colony.

2. Unmated or partially mated daughters, who were sufficiently mature to leave the cell.

Varroa-resistant colonies have, on average, at least twice the level of infertile female mites in the colony (Grindrod and Martin 2021). This has positive implications for mite population control and colony survival.

Fig. 4.6: A high level of infertile foundress mites are found in Varroa-resistant colonies due to bees interrupting their reproduction process.

Resistant colonies have high numbers of infertile foundress mites

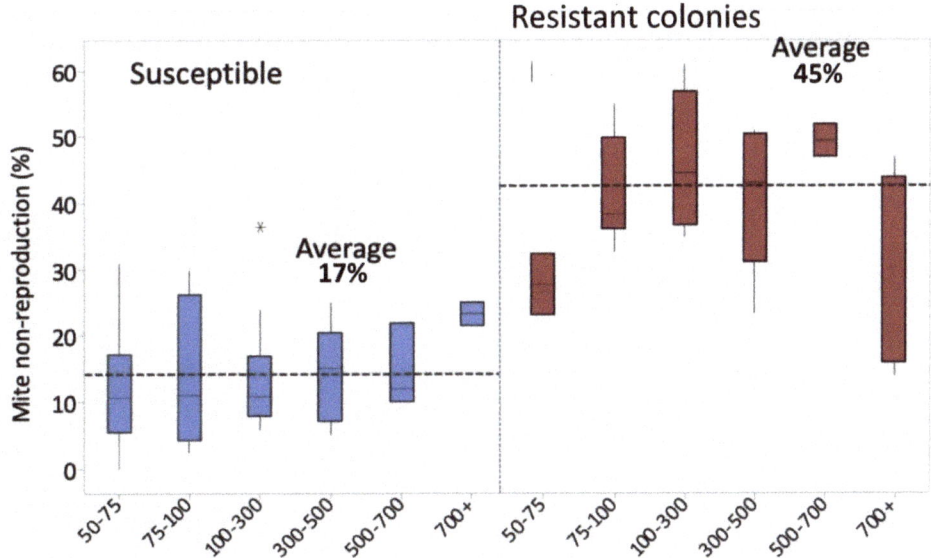

Graph: Grindrod and Martin 2021.

3) Recapping brood cells with healthy pupae

The recapping bees are a different crew. That's their job. They are likely to have lower olfactory sensitivities compared to detector and uncapping bees (Gramacho and Spivak 2003). Recapping bees, finding a cell that is open and unattended, will re-cap it so long as the pupa is alive, even if it is infested with mites.

Recapped healthy pupae still develop normally, as long as the correct humidity and temperature are maintained (Ellis 2010 and Crailsheim 2013). At a beekeeper level, we haven't been able to prove the latter ourselves, but colonies that demonstrate the uncapping / recapping trait are both vigorous in brood rearing and foraging, benefiting from lower levels of Varroa (and therefore vectored virus) through the season.

Detector and uncapping bees, sensing there's still an issue with the recapped cell, can re-open the same cell! It is a reflection of their olfactory abilities and differing roles of nurse bees. Dr. Ralph Büchler observed one cell being opened and re-capped 17 times in an observation hive. Until recently, beekeepers had no idea that this level of activity was going on and it is demonstrated in Chapter 6.

Fig. 4.7: Recapped cells. You can see the darkened centres of these cells where they have been recapped. It can be very difficult or impossible to detect recapped cells in some colonies.

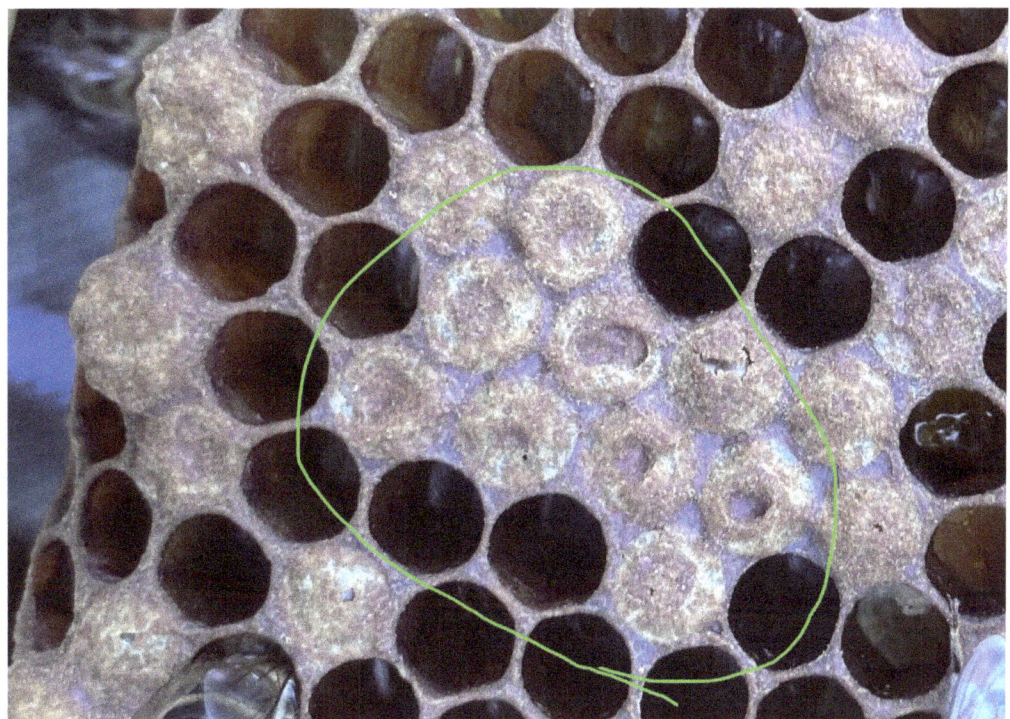

Picture: Steve Riley.

Some researchers focus on the levels of recapped cells in a colony as a proxy for Varroa-resistance. They cut around the cell top and investigate the underside of the cell. A shiny underside base is the undisturbed silk of the cocoon spun by the larva shortly after the cell is capped. But a dark patch in the middle shows where the cell has been uncapped and recapped.

Fig. 4.8: Uncapping a cell to inspect for recapping and a magnified (10x) darkened underside of the cell cap shows where the bees have recapped the surface.

Picture: Dr. T. Rudd.   Picture: Steve Riley.

Our conclusion was that we would not be able to persuade fellow beekeepers to spend time doing this….. and probably sacrificing whole frames of brood to research on. So we trained ourselves to search for the disturbed surfaces of recapped cells. Once you get your eye in, it's not so hard. More on this monitoring approach in the next chapter.

We don't use recapping to monitor hygienic behaviour against Varroa. Uncapped cells are sufficient; they are easy to see, record and a signal that Varroa reproduction interruption is occurring. This is probably the opposite to the rationale for professional researchers, who are not regularly observing the same colonies and need a scientific outcome (e.g. a percentage).

To finish off on recapping is an expansive piece of research (Fig. 4.9), that compares recapping levels of worker brood in known Varroa-resistant *Apis mellifera* colonies, with Varroa susceptible bees, i.e. those with low mite defences and also Varroa naïve bees, where there are no Varroa, e.g. Isle of Man. There are 3 points that jump out:-

1) Recapping (i.e. the uncapping/recapping process) is a strong feature of Varroa-resistant colonies in England and elsewhere in Europe, South Africa, Brazil and even the island of Cuba, where all of their 220,000 colonies of European honey bees (*Apis mellifera*) are treatment free.
2) The level of recapping in susceptible populations is significantly lower, but is still a hygienic trait that bees use (just <u>not</u> sufficiently to control Varroa).
3) The higher red bars show recapping of a Varroa infested cell and yellow bars, a non-infested one, demonstrating that bees have learnt to identify the whereabouts of mites, although sometimes select the wrong cell, as discussed earlier.

Fig. 4.9: Recapping levels of worker brood in Varroa-resistant colonies are consistently high for *Apis mellifera* honey bees across different continents.

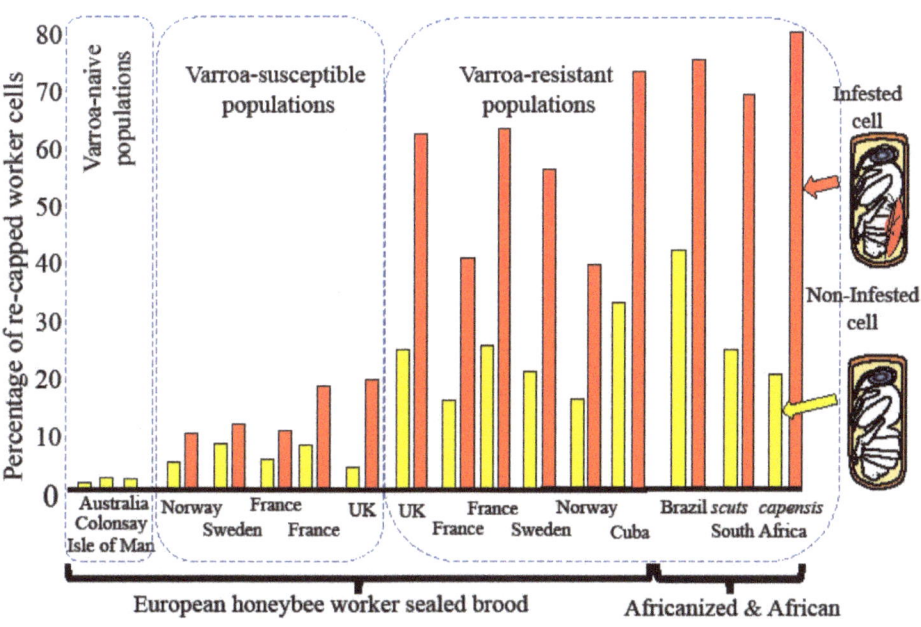

Varroa destructor reproduction and cell re-capping in mite-resistant *Apis mellifera* populations - Martin *et al.* 2020. The European data is taken from Oddie *et al.* 2018, Cuban data from Luis *et al.* 2022, while the Australasian, Colonsay, AHB, and Africa data are from the current study. "scuts" = *A. m. scutellata*; capensis = *A. m. capensis*.

4 How Varroa-resistant colonies manage their mite populations

Isn't this remarkable? Recapping is a strong feature of Varroa-resistant colonies among *Apis mellifera* bees across unconnected continents and different islands, irrespective of the honey bee variety or environment e.g. tropics to northern Europe.

Summary of how resistant honey bees manage their Varroa population

The interruption by honey bees of mite reproduction limits the growth in the Varroa population. With low vectors (i.e. the mites) of virus, DWV is kept at asymptomatic levels. A large cohort of infertile foundress mites is created from exhausting their reproductive capacity through these interruptions.

Fig. 4.10: The diagram illustrates how Varroa and DWV can rapidly increase in honey bee colonies with no or low defences compared to mite population control by bees in resistant colonies with high defences to Varroa.

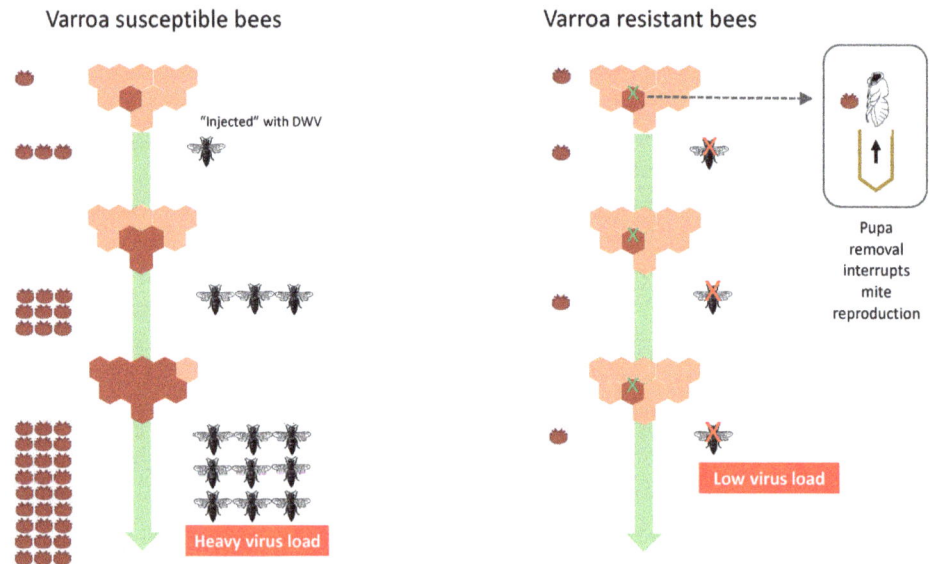

For illustration, 2 daughter mites per breeding cell are assumed to become new foundress mites for the Varroa susceptible bees, which is within the range of estimates of 1.8 in worker cells and 3 for drone cells.

Diagram redrawn with permission and credit to Stephen Martin and team.

In contrast, mite-susceptible colonies with no or low defences to Varroa, allow rapid mite growth, and with it, DWV levels which reduce the lifespan of the bees.

Virus tolerance among Varroa-resistant bees

We know that low mite numbers results in low levels of DWV in a colony. What wasn't appreciated was that virus tolerance occurs in Varroa-resistant bees.

Research led by Barbara Locke and her team at the Swedish University of Agricultural Sciences was able to demonstrate a level of virus tolerance in long-term Varroa-resistant bees on Gotland, versus managed (treated) bees on the same Island. The bees were tested with the same dosage of deformed wing virus (DWV) and acute bee paralysis virus (ABPV), with the result being that the resistant bees lived longer.

The research was expanded across other well-documented Varroa-resistant populations in Norway, the Netherlands and France (Barbara Locke *et al*. 2021). In each case, the results demonstrated that adult bees from naturally resistant colonies are:

….. *much more tolerant to oral DWV or ABPV infection than bees from regular Varroa-susceptible control populations.*

Virus tolerance is an area that requires further research to understand the mechanisms in play. It is unlikely that beekeepers will be able to select for it as a trait, but helpful to know that resistant bees have a viral robustness to them.

5
Why selecting for Varroa-resistant traits is important

Current beekeeping practices are creating further imbalance in the honey bee and Varroa relationship.

Since Varroa appeared in the UK, beekeepers have relied on various chemical miticide treatments and biotechnical methods (i.e. non-chemical) to keep their colonies alive. That was essential when Varroa first arrived since honey bees had low defences to mites and beekeepers had no other strategy for colony survival.

Beekeeper education continues to focus on how the beekeeper can reduce Varroa numbers. There are reams of pages in books, industry magazines, the web and on forums discussing the *how and when* to apply miticide treatments. Over 30 years on, these Varroa reduction methods, are institutionalised in beekeeping and known as Integrated Pest Management.

Now that we are observing honey bees coping with Varroa both in the UK and abroad, the question arises:

Have our past good intentions had unintended consequences?

Mainstream beekeeping does not currently teach about selecting for Varroa-resistance. As a result, the honey bee and Varroa relationship is moving further out of balance. By using miticides, beekeepers are unwittingly removing the natural selection pressure on their honey bees to deal with the mites. The result is a cycle of breeding from bees which are susceptible to Varroa (Fig. 5.1a).

These bees have low mite defences and need treating, or other mite reduction methods to keep them alive – often on multiple occasions through the season.

Some Varroa always survive these treatments and overwinter in the colony. These survivor mites are the breeding stock for new mites in the season ahead. Tough and resilient.

Inadvertently, beekeepers are also selecting for resilient mites, some of which have become resistant to chemical treatments. Over or incorrect use of miticides led to mite resistant issues from pyrethroids such as Apistan® and Bayvarol® seen in 2001-02 and more recently from Apivar®, where amitraz is the active ingredient. In beekeeping forums, recommendations for treatments include multiple applications of oxalic acid or combinations of treatments with other miticides over the season. Back in the 1990's, one dose of Apistan® or Bayvarol® in late summer would suffice. More miticide treatments are now required to kill resilient mites and keep susceptible bees alive.

Beekeepers are perpetuating the Varroa problem by <u>not</u> selecting for Varroa-resistant traits in honey bees and increasing the selection pressure on the mites to be resilient.

Fig. 5.1a: Current cycle of beekeeping perpetuates bees' susceptibility to Varroa.

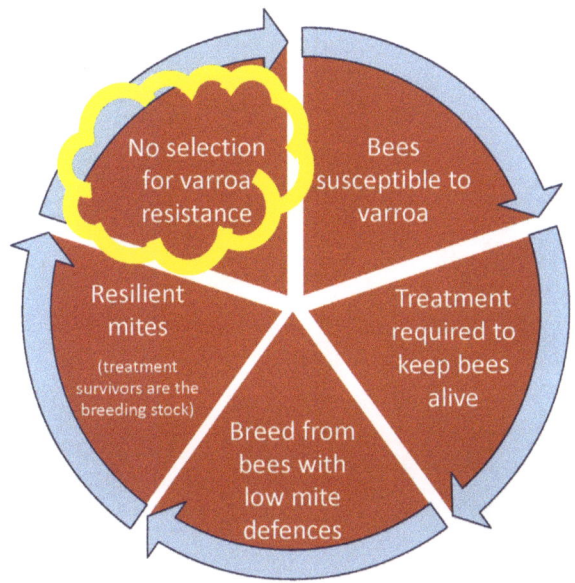

Chart: Steve Riley.

Selecting for Varroa-resistance stabilises the honey bee / mite relationship

The whole cycle changes when the beekeeper introduces Varroa-resistance into the selection process (Fig. 5.1b). No treatments are required, after a period of selection. Colony divisions, spare queens and drones spread resistant traits into the local vicinity.

For the beekeeper, there are big savings in time and money when mite-treatments become unnecessary. There is also flexibility of when to take-off surplus honey.

If our climate continues to warm, we should expect brood rearing and mite reproduction to continue later into the year. Resistant traits are an insurance against late season mite growth.

Fig. 5.1b: Selecting for Varroa-resistance means identifying bees with defences against mites that control their population. These bees do not require miticide treatments to keep them alive.

Resistant bees restrict Varroa from freely reproducing in worker brood resulting in lower mite and virus levels, and a host/parasite equilibrium. The bees' traits proliferate locally and embed in the locality.

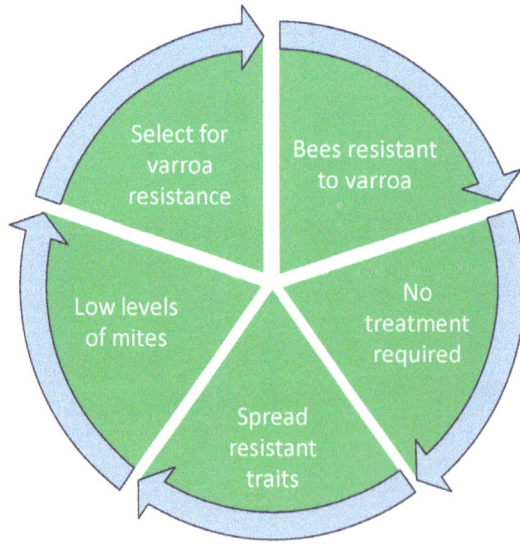

Chart: Steve Riley.

6
Westerham Beekeepers' monitoring of Varroa-resistant traits

The Westerham Beekeepers' project started in 2017 inspired by reports of honey bees surviving Varroa in the UK and different parts of the world. There were beekeepers in the UK who hadn't treated for many years, but the mechanisms the bees were deploying to control their mite populations were not fully understood.

In the densely populated south-east of England, there were few advantages from large numbers of long-lived naturally resistant wild or unmanaged colonies, where Varroa-resistance had developed from natural selection. We recognised the need for a scientific underpin to our approach and read many of the open access research papers on naturally resistant colonies, and also on the Varroa-resistance project being run by the United States Department of Agriculture ("USDA"), including early research on uncapping and the suppressing of mite reproduction (Harbo and Harris 2005). We had also met Dr. Ralph Büchler at Gormanston beekeeping summer school in Ireland in July 2017, who was to play an important role in the early years of our project.

We put a lot of effort into researching our colonies to aid our understanding and facilitate knowledge-sharing locally. There was an initial investment in beekeeper time to monitor for resistant traits, with the trade-off being time and money saved by not applying miticide treatments.

The section that follows is the end result of what was an evolving monitoring process that started with assessing mite levels. Hopefully the detail will prove to be useful reference material for fellow beekeepers and facilitate accelerated learning.

Monitoring

The objective of our monitoring was for the beekeeper to build a view of a colony's ability to manage Varroa. *Does the colony have HIGH or LOW defences against mites?*

The approach at Westerham Beekeepers was to observe the bees in their interactions with Varroa. Naturally resistant colonies 'knew' the traits and we needed a framework to identify and monitor them in our own hives. We eschewed complexity, the enemy of success. It had to be straight forward and communicable around a beekeeper group.

We saw monitoring as an extension of the inspection regime and sought to carry it out in the least invasive way possible. The rest of this chapter covers our approach to monitoring:-

- Mite levels
- Uncapping / recapping
- Chewed-out pupae

Monitoring mite levels - counting mites from natural drop

This is not a conversation starter at a party. Even beekeeper colleagues roll their eyes. But this magnificently dull task produces insights into what's going on in the colony, including bonus research findings from honey bees' interaction with Varroa. The humble mite board under an open mesh floor ("OMF") enables the beekeeper to track the mechanism that bees use to control mites from January to December, without opening the hive. An invaluable and underestimated tool.

Open mesh, sometimes referred to as screened floors became the standard floor offering after Varroa arrived; the rationale being that Varroa would fall through and not be able to re-enter the colony – a biotechnical reduction. OMFs were trying to reduce a problem rather than solve it. Honey bees typically don't like open-sided nests and certainly don't need the ventilation or draughts. When you insert a Varroa board under an OMF, you have a research lab generating a wealth of information. Plus a quasi-floor which, in our view, is better than an open one.

Fig. 6.1: A floor base with an open mesh, where mites and chewed-out pupae exoskeleton fall through onto a slide-in insert board (not shown) underneath, providing a wealth of information.

Picture: Steve Riley.

Monitoring mite drop needs interpretation and correlating with the seasonal priorities and brood in the colony. See it as part of the re-engagement by the beekeeper in the interaction between the honey bee and Varroa. For us, the insert board provided the following valuable insights:-

1) Varroa data through the season as a guide to the level of the underlying mite population in the colony and comparison between colonies. Without data you are guessing blindly.
2) Monitoring pupal exoskeleton on the insert board gave us valuable insights, many not recorded by scientific research, including:

i. Evidence from January onwards after the first of the new season brood has been laid. The beekeeper knows that the honey bees' hygienic traits against Varroa have been triggered following the overwintering foundress mites entering the brood cells to reproduce. Varroa reproduction is already being interrupted and unhealthy pupae are being evicted.

ii. Understanding that chewing-out pupal exoskeleton is an integral component of Varroa-resistance where the Varroa offspring die. Without this trait, we find mite numbers climb to dangerous levels.

iii. Hygienic behaviour against mites fluctuates dependent on the strength of the nectar flow. Fewer pupal exoskeleton of workers are seen during the spring flow, partly as Varroa are diverted towards drone brood. Also true during a strong summer flow, when the colony prioritises nectar processing for winter stores over hygienic behaviour. This reverts back as the flow declines and ceases.

iv. Informs the beekeeper that the queen is alive and started laying some 15-17 days earlier (i.e. 3 days an egg, 6 days a larva then chewed-out 6-8 days after the cell was sealed). It is cheering to see this in the winter.

3) Evidence of a mass pre-winter clean-out of mites in the late summer is very visible using insert boards, as discussed in Chapter 8, providing an opportunity to understand rather than fear this colony behaviour.

As an aside, many of these important learnings would have been missed counting mites in a 300-bee sample method (e.g. alcohol washes or sugar shakes).

Methodology

The author's Varroa boards are inspected every 2 or 3 days, then cleaned and re-inserted. This is made easier by having a garden-apiary.

One of the founder group at Westerham Beekeepers puts a board in for 2-3 days a month. Taking 12 readings a year will provide a sufficient guide. It is only a view of HIGH or LOW mite levels that is required.

Another beekeeper reviews their boards weekly, where there will be more hive debris to sort through and a magnifying glass or blowing up pictures from a phone is helpful.

A brief record is taken of:-

- the number of adult foundress mites (dark brown/red); these are key as they are capable of reproduction (Fig. 6.2).
- juvenile daughters (light tan to white) or sons (white/slightly yellow); a high proportion informs that bees are cleaning out cells and large numbers are often seen in late summer as brood numbers reduce (see analysis in Chapter 8).
- any white pupae exoskeleton (antennae, head and thorax typically) and the amount.
- anything else of note, which might include wax scales, ejected eggs, pollen stripped by the OMF, position of the cluster as it moves around through the winter, or number of frames the bees are covering judged by the wax clippings from brood or bees accessing honey. A picture of colony behaviour can be deduced without opening the hive.

Fig. 6.2: Only adult (foundress) mites are counted as these have reproductive capacity (although up to half of them are infertile in resistant colonies).

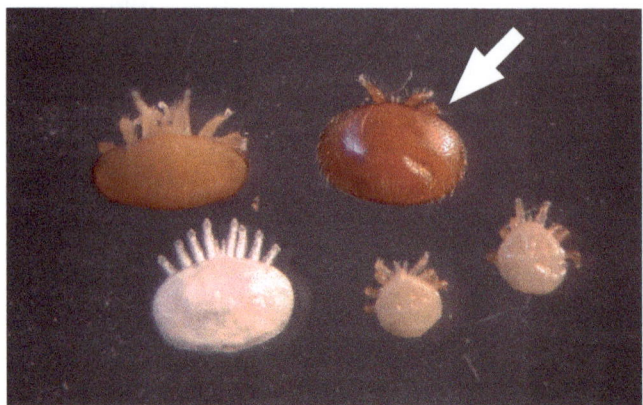

Picture: © Zachary Huang Photo.

We have recorded tens of thousands of mites – you get efficient and skilled at interpreting developments inside the colony, without disturbing the bees' homeostasis. The notes of mite numbers are transferred to a spreadsheet, where graphs of trends can be seen, correlated with brood in the colonies. Interpretation of the data drives the selection of queens and more on this in Chapter 10.

Fig. 6.3: The humble insert board, as well as being your *research lab* for Varroa and chewed-out pupal exoskeleton, will acquaint you with many other visitors to the hive. Pollen mites, springtails, woodlice, earwigs, wax moth, ants, leopard slugs will all sneak through a small gap. Natural visitors of a colony. To keep monitoring quick and simple, we do not add any adhesive such as Vaseline to the board and prefer the boards with a lip at the back to keep out draughts.

Picture: Steve Riley.

How many mites is ok?

In the initial phase of evaluating colonies for resistant traits, it is important to monitor mite counts over time (minimum of monthly) but not in isolation – include observing for uncapping plus chewed-out pupae.

Across an entire year, we like to see an average mite drop of under or about 5 per day for our best colonies. This is c.1,825 over the year. A typical profile will be low drop in the early to mid-season, where there is brood, with a peak fall in autumn. Seasonal fluctuations occur with brood and nectar flows, which are discussed in the next chapter.

Fig. 6.4: This is our mite drop guide for a British National hive, with the brood area restricted to the brood box. Scale the guidance according to hive size and queen fecundity – high laying queens will have more brood and mites in proportion.

Average daily (adult) mite drop	Colony risk	Action
≤ 5 per day (for the year)	Low	Breeder
6-10 per day "	Medium	Judgement required
>10 per day "	High	Requeen

In the early years of our project, mite drop levels were towards the medium to high range, but have gravitated down, benefiting from selection, requeening and some colony losses of those with the weakest traits.

These mite drop numbers are broadly comparable with Varroa levels gathered from a research project in France (Le Conte *et al.* 2007). Unmanaged colonies were discovered to be surviving Varroa without beekeeper management in Le Mans and Avignon. Mite counts were taken one to three times a week using natural mite drop to a screened bottom board (same method as us) for a full year from September 2002 to August 2003. In 12 resistant colonies, the average mite drop for a year, put onto a daily basis, was 9.1 per day. These were for Dadant hives which have a brood area c.70% larger than a British National. Adjusting for the larger brood area, the equivalent daily mite drop for a British National is 6.4 per day, similar to our target.

Monitoring: Uncapping / recapping

Uncapped pupae at the pinkish-eyed stage are easy to see during inspections. As the queen lays in age-groups, nearby cells are worth scrutinising for the start or finish of uncapping (you can never tell which). If you have a really good eye, that is where the disturbed surface of recapped cells can also be found.

We record uncapping during inspections on a hive record card. It becomes part of the beekeepers' normal regime. With monitoring for Varroa-resistant traits, you quickly build up a knowledge of which colonies have them. In turn, this feeds into queen rearing decisions.

Investigating worker cell cappings with bees flying around the veil can be tricky. Taking pictures of brood frames to blow up and analyse from the comfort of the kitchen table reveals a lot more than we ever appreciated. Having spotted some uncapped cells, move the bees from the area and blast off some quick pictures. You should be dealing with nurse bees so they will be amenable.

For a more complete investigation, take a picture of a full frame. This involves checking that the queen isn't on the frame first before shaking off the bees into the hive box. Get the sun straight onto the frame before taking pictures. Return the frame as quickly as possible. This exercise is much easier to do with two of you.

Fig. 6.5: Reading a brood frame for resistant traits. The picture shows a section of worker brood from a Varroa-resistant colony of Dr. T. Rudd. The queen has most recently laid at the bottom of the frame where there are eggs and larvae. The next section up includes recently capped cells, which are smooth and convex (not so easily seen in print). Then there are a group of uncapped cells, with a lip of wax capping rolled-back around the cell edges ready to be reused. The bees have uncapped these about 6 days after the cells were sealed, which is ahead of the brother and sister mites reproducing. Finally, a little further up the brood frame are the disturbed cell cap surfaces of where they have been uncapped then recapped.

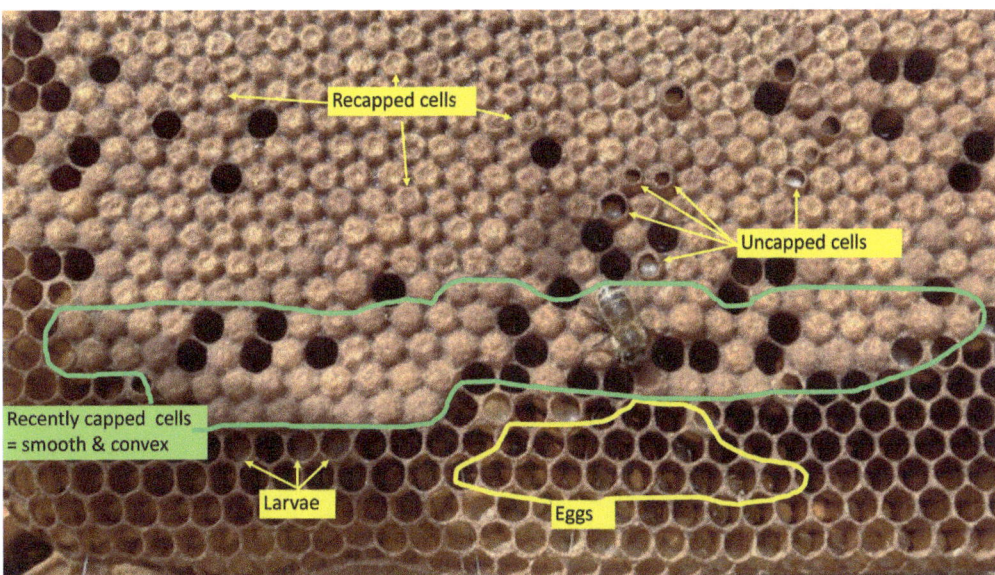

Picture: Dr. T. Rudd.

Monitoring: Chewing-out infected pupae

We came to learn that this was an important trait for beekeepers to monitor. Initially the excitement was around uncapping, as the little white faces of pupae were easy to spot. But that was just the start of the investigatory process by the bees. The chewing-out of pupae from an infested cell was the completion of that investigation. First out would be the antennae, followed by white exoskeleton of the head and thorax. These fall through the open mesh floor onto the insert board and are easily visible when monitoring for mite drop. Immature mite family members would also be cleaned out by the bees and can be seen on the insert board, although the mother mite escapes. A magnifying glass or taking a picture and blowing it up later can be useful.

Fig. 6.6: Antennae and white exoskeleton on the insert board confirm the chewing-out process.

Picture: Steve Riley.

We have never witnessed the abdomens of pupae being chewed-down by bees and don't see them on the insert board. On warm days, slow flying undertaker bees have been observed with what looks like a small white flag. This is the white abdomen of a pupa being flown away from the nest for hygienic purposes. In the cold of the winter, they are dumped outside the entrance and then disappear, perhaps on a milder flying day taken by a bee or as an easy snack for a passing robin.

Fig. 6.7: Pupal bits dumped outside of the entrance on milder winter days.

Picture: Steve Riley.

Monitoring Varroa-resistant traits over 24/48 hours

This was not part of our regular monitoring but one-off research to improve our understanding of the level of hygienic behaviour against Varroa going on inside our colonies. It involved Hive 6 in the author's apiary, where pictures were taken of the same section of brood comb 24 hours apart. Disturbing a colony twice in 2 days felt a little uncomfortable, but we would recommend the approach if a beekeeper needed confirmatory evidence of a colony's Varroa ability, perhaps as their breeder queen.

The findings were illuminating.

Hive 6 had a record of low mite counts, with observed uncapping from inspections and pupae bits on the insert board.

On the first day (Monday) of the brood-frame inspection, there was plenty of uncapping evident at the pink-to-purple-eyed stage. To the left of the uncapped cells in Fig. 6.8, there were also recapped cells, which are not easy to distinguish in the photo.

On the Tuesday, after 24 hours, there had been further uncapping, lots of chewed-out pupae and one cell being recapped (and a few more next to that behind the awkwardly positioned bee).

Fig. 6.8: Shows the change in uncapped and recapped cells in 24 hours. Plus a number of cells where the pupae have been chewed-out.

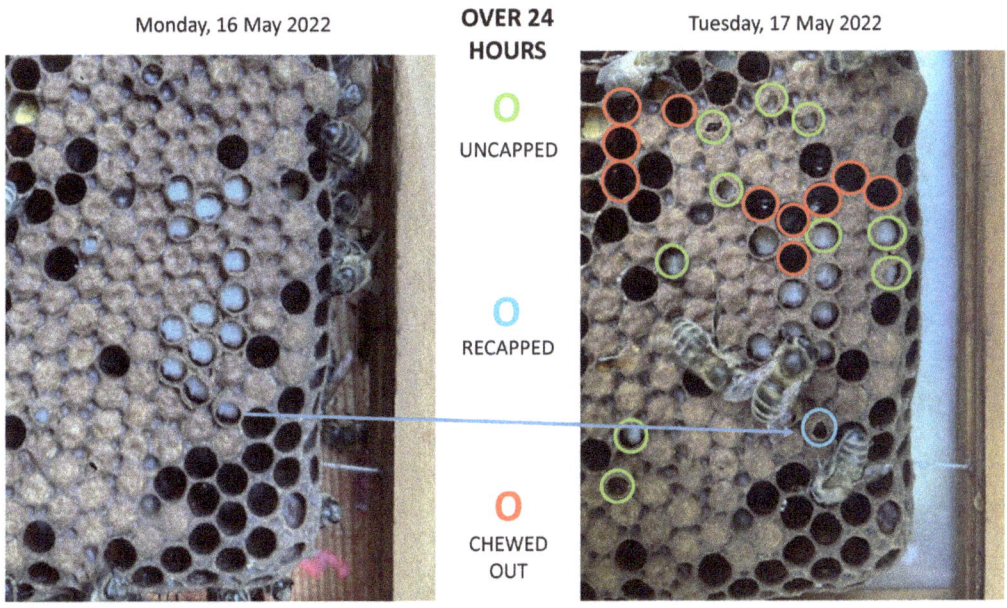

Pictures: Steve Riley.

If we combine the total activity from Monday together with Tuesday, a remarkable level of hygienic behaviour is evident over the 2 days (see Fig. 6.9). It surprised us, but is consistent with the high level of recapping (proxy for hygienic behaviour) recorded across continents where *Apis mellifera* colonies have become naturally resistant to Varroa shown in Fig. 4.9.

Fig. 6.9: Combining the 2-day activity confirmed the high level of hygienic behaviour against Varroa.

Monday & Tuesday, 16-17 May 2022

O UNCAPPED

O RECAPPED

O CHEWED OUT

Picture: Steve Riley.

The take-away for beekeepers is to monitor. There is more bee-led activity going on against Varroa than is appreciated.

Summary: sufficiency of hygienic behaviour is key

It is important that there is <u>sufficient</u> hygienic behaviour by the bees against Varroa to keep mite and therefore DWV at low levels, i.e. sufficient uncapping and chewing-out of worker brood to interrupt Varroa reproduction. The answer is corroborated by a low level of mites in the colony through the season (aiming for 5 or under a day on average for the year).

Monitoring the insert board for mites and pupae exoskeleton, looking for uncapping during inspections and the occasional photo of brood combs will allow the beekeeper to identify Varroa-resistant traits in their colonies.

Fig. 6.10: Beekeeper monitoring for Varroa-resistant traits is a 3-step process.

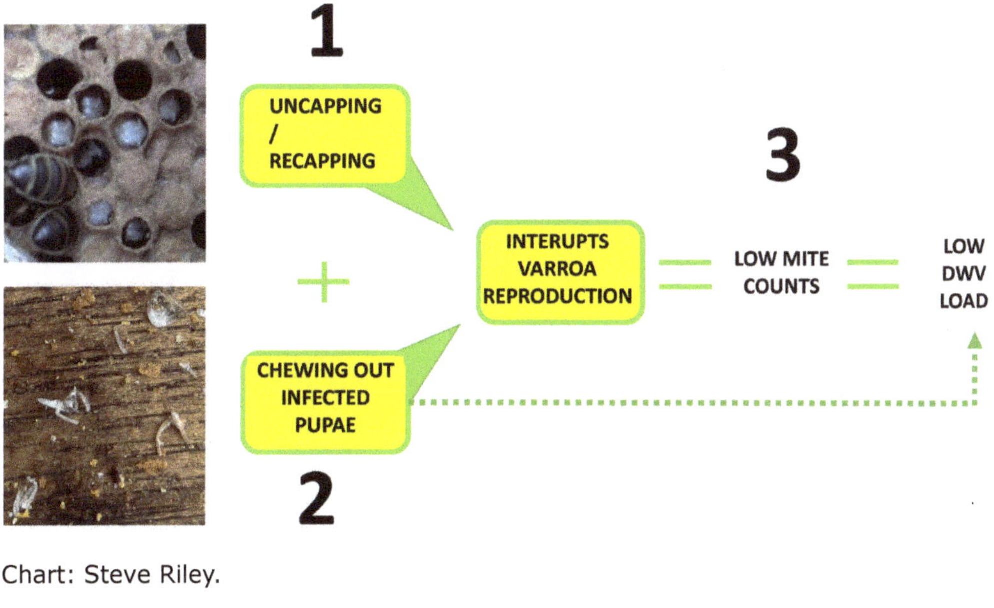

Chart: Steve Riley.

7
Other approaches to monitoring

Whilst we adopted the Varroa insert floor and mite drop method to monitor for resistant traits, there are other approaches to determining a colony's ability to deal with Varroa. Some of these we considered and dismissed, believing that our beekeepers wouldn't adopt them or that they were overly invasive or unproven as yet.

These methods included:

- Sampling e.g. sugar shake, CO_2 or alcohol / detergent wash.
- Brood analysis e.g. VSH test, drone brood uncapping.
- Tests for hygienic behaviours e.g. pin-prick test or UBOs.

Summary comments on these approaches follow, starting with a sampling method using icing sugar.

Sugar shake

At our Training Apiary, we had trialled the sugar shake to estimate mite loads. This was one of the methods deployed by our then mentor, Dr. Ralph Büchler, who advised that *'as long as the sugar is dry, it will be almost as good as an alcohol wash'*.

Method: nurse bees from a brood frame are bumped into an empty shaker pot (not the queen!) with a 100 ml line drawn around it (N.B. pour in a measure of 100 ml of water earlier to get the line level); this is equivalent to about 300 bees. Two tablespoons of icing sugar are added. The bees are slowly turned in a tumble-dryer effect inside the pot, rested and tumbled again. Finally, they are shaken over a white piece of paper until mites fall through the mesh in the lid of the pot. About 15 minutes later, you have a mite count and the powdered white, sugar dusted, rather dizzy bees are returned to the colony.

Fig. 7.1: Sugar shake with white dusted bees.

Picture: Steve Riley.

This is a crude estimate of infestation. Only mites on 300 bees are counted. Making a judgement by extrapolating an infestation rate of Varroa in a calculation using under 1% of the bees in a colony of over 30,000+ is open to misinterpretation.

The attraction of sampling is a clean percentage infestation rate. Finding 9 mites would give you a number of 3% after adjusting for the c.300 bees. Should one correct for an estimated 50% - 80% of mites reproducing in sealed brood?

Research on Varroa-resistant colonies in the UK showed an <u>average</u> infestation rate of 6% of infested adult workers (Grindrod and Martin 2021). Any colony with over 10% of infested adult workers in September should be regarded as high risk with weak traits. At other times of the year, you would expect lower infestation levels.

Sampling can only really take place during the inspection season of March to September. What happens to Varroa levels and the interaction with bees later in the autumn, winter and early spring? We just had to know!

Monitoring Varroa from a small sample of bees wasn't for us, which also meant avoiding a CO_2 test, alcohol and detergent wash, where for the latter techniques, some 300 bees are killed per test on each colony. Extrapolating mite infestation rates from small samples of bees gives an illusion of accuracy, but the reality can be a large range of infestation due to the non-monitoring of mites in brood.

Varroa Sensitive Hygiene ("VSH")

Some commercial breeders test for Varroa Sensitive Hygiene. This was first developed as an approach at the USDA Honey Bee Laboratory in Baton Rouge, Louisiana. It involves opening 100 sealed cells of worker pupae at the purple-eyed stage and counting Varroa with or without offspring. Practitioners of this approach include Dr. John Harbo, who led the original research (https://www.harbobeeco.com/) and who includes a free assay available from his website, and breeder Cory Stevens (https://www.stevensbeeco.com/), both in the USA.

Stevens uses the Harbo assay for selection and believes that establishing resistant bees with this approach could take between 3 and 10 years. The shortest timescale would depend on the traits of the starter queen, whether other VSH queens are brought into the programme every year, the de-selection of any queens not expressing the required level of VSH traits and using instrumental insemination from selected drones, i.e. a closed population breeding model.

Using the Harbo assay on unselected colonies (i.e. starting without a VSH queen) could require the longer period to establish VSH traits.

At Westerham Beekeepers, we considered adopting VSH brood analysis, but couldn't envisage enough of our beekeepers spending time analysing (and sacrificing) worker brood combs from their colonies.

Drone brood uncapping

This approach involves the uncapping of 100 live drone pupae at the purple-eyed stage. Finding 5-10 mites in 100 is purported to represent severe risk to the colony, according to the National Bee Unit in their leaflet, "Managing Varroa."

This is not a practice we would recommend as it contributes towards poor queen matings and is an unpleasant monitoring method. Inevitably, one accidentally forks out recently capped pre-pupae creating a mushy mess.

The key battleground for Varroa is in worker brood and drones are not always available to fork out. Consistency in approach is important for comparability.

Fig. 7.2: Forking out live drone pupae is mainstream teaching to monitor for Varroa; it is a practice that contributes towards poor queen mating, especially in the second half of the season and is, in our view, unnecessarily barbaric.

Picture: Courtesy of The Animal and Plant Health Agency (APHA), Crown Copyright.

Pin-prick or freeze-kill method

Earlier efforts by breeders and researchers, including at the University of Sussex, to identify Varroa-resistance had used the pin-prick test or freeze-kill brood method to identify hygienic behaviour. A small patch of brood would be killed or damaged and then checked either 24 or 48 hours later to judge how many of the pupae had been removed by the bees. The tests had flaws:-

1. Varroa rarely kills the pupae in the cell and breeders were selecting for bees that could detect large groups of dead brood (chemical cues from dead and live brood are different).

2. There was no test for Varroa infestation in the cell.

Unhealthy Brood Odours ("UBO"s)

Research and field tests are underway for UBOs. These are sprayed onto sealed worker brood cells and designed to quickly test for bees' hygienic behaviour against Varroa, judged by how many cells are uncapped in two hours. Getting the correct chemical cues is complex and it will be interesting to see if there is a strong correlation to the practices of Varroa-resistant bees after sufficient field trials with beekeepers – a minimum of 3 years of not treating will be required in our view.

At the Westerham Beekeepers' project, we are allowing the bees to tell us whether they have the detector ability to sense Varroa infestation in the cells, through observing for uncapping and chewing-out of pupae. Honey bee-led science.

Genetic markers

Other researchers at the USDA have looked for genetic markers linked to bees' resistant traits to aid breeding. These have eluded them with low agreement over identifying the correct genes. This is a complex area not helped by the polyandrous mating of virgin queens and the bees' high genetic recombination rate (ability to mix up genes to turn on and off different traits).

Summary – monitoring

In the wrap-up on monitoring, the Westerham Beekeeper approach has been included to highlight its merits.

Varroa insert board:

Our chosen method for monitoring was mite drop using an insert board under an open mesh floor. To us it had many advantages:-

- It is a test of mites from the whole colony (not a small sample)
- Identifies chewed-out pupae exoskeleton (important part of the resistance process).
- Is non-invasive.
- Easy and quick to use.
- Allows monitoring from January to December.
- Provides insights into the seasonality of hygienic behaviour.
- Whilst providing a floor for the colony.

There are two main criticisms:-

Fears that mites are being eaten or carried off by ants ruining the counts. We have never observed that, but losing a few when counting from a sample of 30,000-40,000 bees should not be a concern, i.e. the maths are not affected much.

Counts do not reflect the build-up of Varroa in brood – this is true to a degree (more so for sampling methods), but boards will show an uptick in Varroa numbers when there are hatches of bees and mites emerging. The key is to monitor over time, which will show the trends and differential Varroa abilities between colonies.

Brood monitoring from pictures:

Taking pictures of worker brood having cleared them of bees (watch out for the queen) and magnifying the picture later in the sanctity of the kitchen, will show the level of uncapping that's going on – sometimes the disturbed surface of the cells reveals recapping as well.

As discussed in the previous chapter, a picture of the same brood comb 24 hours apart can reveal activity of uncapping, recapping and chewing-out of pupae. A level of disruption not to impose on a colony more than once a season.

Monitoring - key points

There are imperfections or drawbacks for all Varroa monitoring methods. Whichever process is selected, use the results as a starting point to identify Varroa resistant traits:-

1. Make sure monitoring is carried out regularly (minimum of monthly) and not once a year.
2. Be wary of drawing firm conclusions from small point-in-time sampling methods which test for Varroa on bees and ignores mites in brood.
3. Use the same method consistently for comparable results.
4. Correlate your Varroa level evaluation with observing uncapping and chewed-out pupae – key parts of the resistance process.
5. Integrate monitoring into hive records to drive queen selection / de-selection.
6. Build a picture of your different colonies' strength with regards to Varroa and rank them.

8
Seasonality of honey bees' hygienic behaviour against Varroa

This chapter takes a deeper look at honey bees' hygienic behaviour against mites through a full season. The commentary covers a Varroa-resistant colony, now in its seventh season, with observations on the effect of drones, swarming, nectar flows and pre-winter preparations.

In the first section, as a reference guide for beekeepers, a whole year's mite drop has been recorded and correlated with brood rearing. Interpretations have been made through the year of the interaction between bees and mites (Fig 8.1b).

In the second section, using the full-year mite drop data as a base, an estimate of how actual Varroa levels change through the season has been profiled together with the key drivers (Fig. 8.8).

Both sections have been divided up into January to March, April to May, June to July, August to October and November to December. The aim here is to consider the honey bee and Varroa interaction during key seasonal events for the colony.

Set up of the hive being monitored

This commentary is best read whilst referencing the full year analysis of Hive 1 during 2021 (Fig. 8.1b). This a Varroa-resistant colony in the author's apiary. The bees have been locally adapted from this apiary for over 10 years and have darkened over this period. They attempt a reproductive swarm every year and the daughter queens have the same or very similar Varroa-resistant traits as their mothers (N.B. good heritability). Hive 1 has the best track record in the apiary for producing surplus honey and has never suffered a loss over winter. The bees are pleasant to handle (most of the time).

Fig. 8.1a: Hive 1; a fully Varroa-resistant colony. The set up is a British National brood box, queen excluder with an insert board in 24/7 under the open mesh floor, also acting as a quasi-floor.

Picture: Steve Riley.

Seasonality of mite drop (using Fig. 8.1b)

January- February- March:

In January, the first of the new worker brood has already been laid (starts in December where we are). Small amounts, the size of a palm of one's hand. For the Varroa, which have successfully overwintered on the bees, this is their opportunity to reproduce.

Interestingly, earlier research on winter colonies showed a high level of male mite mortality in the winter (42% versus 18% summer) leading to infertility among female mites, who depend on their brother for mating (Martin 2001).

Mite drop, which is just a trickle (1-5 per day), should trend down initially as mites move into worker brood. Slightly counterintuitive: new brood = lower mite drop.

Fig 8.1b: Mite drop profile for Hive 1 during 2021, correlated with brood and bees' hygienic behaviour against Varroa.

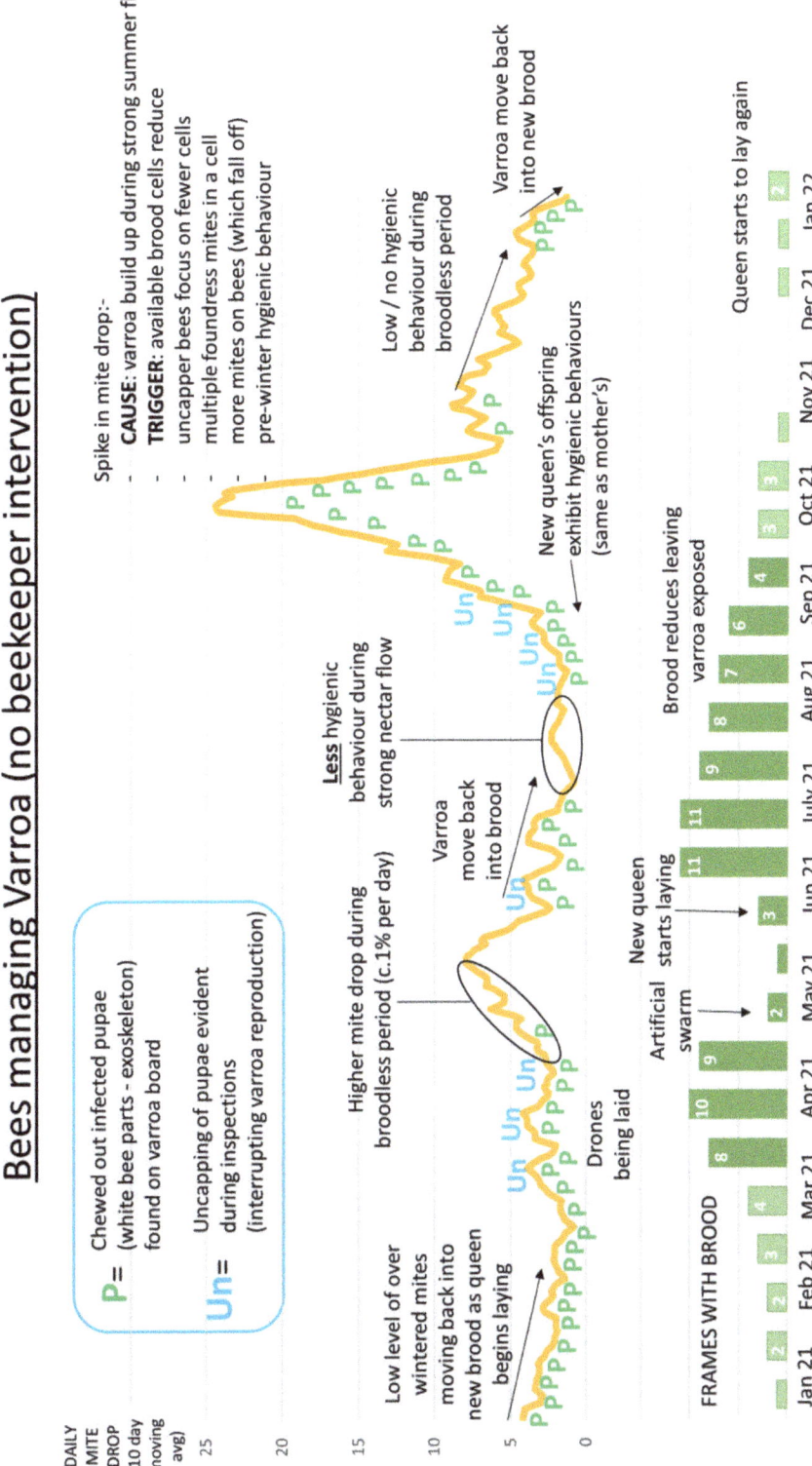

Chart: Steve Riley.

8 Seasonality of honey bees' hygienic behaviour against Varroa

Sixteen to seventeen days after the queen starts laying (pupae now at the pink-to-purple-eyed stage), the bees chew-out infested pupae. The beekeeper can observe white exoskeleton on the insert board. A joyous moment. Firstly, the beekeeper knows that the queen is alive and laying (hopefully seeing worker not drone exoskeleton!) and secondly, that the bees will deal with overwintering Varroa at an early stage in the new season, in contrast to the build-up seen in colonies with low defences to Varroa.

As well as monitoring for hygienic behaviour, the insert board is a non-invasive provider of other information. During milder spells of weather, small droplets of liquid honey or crystals from ivy honey show up across a wide area of the insert board as the colony accesses their stores. Small bits of pollen from Mahonia, *Fatsia japonica* or winter-flowering viburnum can sometimes be seen in December or January – always a good sign.

In the colder snaps of winter, expect the wax and cell clippings to reduce over a smaller area of the insert board, as the cluster tightens.

Fig. 8.2: Wax clippings from a cluster during cold weather in January over 3 days, which shows the position of the colony over 5-6 frames. The cluster movements can be tracked over winter as the colony accesses honey stores. On this occasion, there were 5 mites (i.e. 1.7 per day).

Picture: Steve Riley.

What we like to see in late January is evidence on the insert boards of bees clustered across 5-6 frames, working on keeping the first brood warm.

April-May:

Poor weather

When bees are kept in the hive due to heavy rain or cooler conditions, hygienic behaviour against Varroa rises, observed by higher pupae exoskeleton on the Varroa board. The nurse bees are not distracted by receiving and processing nectar or packing down pollen.

Drones

Ahead of the reproductive (swarm) season, the first drones are laid towards the end of March and into April. These are over 8x more attractive for Varroa, than worker brood (Fuchs 1990).

At a colony level, the drones' sacrifice relieves mite pressure on worker brood, leaving them to develop strongly, which supports foraging of spring pollen and nectar flow, building towards the reproductive swarm season.

As an aside, we do see some uncapping of drone brood by our most hygienic colonies where mite levels are already low. To us, this is the start of the same process of Varroa reproduction interruption as occurs in worker brood, with the bees sensing Varroa offspring or infected pupae. Currently, research hasn't been able to prove this hygienic behaviour in drone cells is followed through to interrupting Varroa reproduction.

Fig. 8.3: Drone brood chewed-out on the insert board – the large eyes confirm the identity.

Picture: Steve Riley.

There are two observational consequences of the drone brood presence for the monitoring beekeeper. Firstly, there will be <u>less</u> uncapping/recapping of worker brood and chewed-out pupae, with mites attracted to drone cells. And secondly, there will be a blip upwards in the mite drop as the drones hatch after 24 days. The mite population will grow due to the longer pupation period of 14 versus 12 days and less (or no) hygienic activity by the bees against the mites in the sealed drone brood. But there is a neat offsetting synergy here with swarming, which comes next.

Swarming and brood-breaks

The spring flow in our area, supports strong brood development of workers and drones in the colony and we usually see the first signs of swarm intentions from late April onwards. We don't try to thwart swarm preparations, but anticipate and work with the timing of the bees using the nuc artificial swarm method.

Post swarming, the subsequent broodless period, before the new queen begins to lay, forces the Varroa onto the bees and an increase in mite drop will be seen from accidentally falling (possibly groomed?). This is a helpful feature of every broodless period. In this case from Hive 1 in Fig. 8.1b, about 170 mites dropped in total during this break in brood, where mite drop increased from 3 to 8 per day at the peak.

Dr. Ralph Büchler estimates that mite population drops about 1% each day during broodless periods (Gormanston lectures 2017). A broodless period will vary depending on when the old queen stopped laying and when the new queen gets mated and re-starts brood rearing. Assuming a range of 20-25 days (can be longer if poor weather prevents the new queen being mated) when Varroa are forced onto the bees, this could result in a decrease in the mite population by about one fifth to a quarter. This is a helpful offset following growth in the mite population from successful reproduction in drone brood.

June-July:

If there has been a swarm event in late spring, then hopefully the new queen is well mated and will be laying strongly by now. Mites which had survived on the bees through the broodless period are now seeking opportunities to reproduce in the new worker brood. Consequently, mite drop numbers on the insert board will decline.

The end of June into July typically sees the strongest part of the summer flow in our area, fuelled by blackberry with clover and Himalayan balsam. These are the essential winter stores for the colony. Foragers are working from dawn to dusk bringing back large volumes of nectar to be received by younger bees in the hive, who have responsibility to add appropriate enzymes, reduce the water content and wax-seal it as honey. The colony is in 'honey factory' mode and hygienic behaviour against mites becomes a subordinated activity, given this focus to bring in and process winter stores.

As a result of the nurse bees' focus on receiving and processing stores, the beekeeper will observe <u>less</u> uncapping/recapping in the brood nest and chewed-out pupae bits on the insert board. Consequently, there will be growth in the mite population who breed with less intervention by the distracted nurse bees.

Around this time of year, mainstream teaching encourages beekeepers to assess their mite levels, with a view to drawing a conclusion on whether to treat or not. This is too late in the season. Varroa levels will be underestimated at this time on insert boards, due to reduced interrupted mite rearing in worker brood. Sampling methods such as an alcohol wash or sugar shake will also not show the growth of mites in brood. Analysing a 100 cells of pink-to-purple-eyed worker brood would provide the clearest indication of infestation levels, but this is a route we decided not to go down or could not envisage enough beekeepers adopting, at a time when they are preoccupied with putting on and taking off honey supers. We settled for understanding what was going on between mites and bees in the height of summer.

Good husbandry is to assess Varroa levels once a month at a minimum and gain an understanding of the interactions between brood availability and hygienic behaviour against the mites.

August-September-October:

Until now, the bees are focused on securing sufficient stores to overwinter. As the summer flow declines, their attention turns to readying the nest for winter. This includes the laying down of large amounts of propolis, witnessed by most beekeepers in their final inspections of the season. Propolis was once seen as a nuisance by beekeepers but with its anti-microbial properties, we are now starting to realise how important it is for the health of our bees.

"Cerro Torre" peak – the pre-winter spike in mite drop

What interested us in our naturally resistant colonies was the mass mortality clean-out of mites that occurs in late summer. The mite drop during this period in the colony shown in Fig. 8.1b amounted to 661; that's about 50% of the mite drop for the whole season. The bees were naturally achieving a reduction in Varroa akin to the application of a miticide treatment over the period. With this reduction in Varroa goes a commensurate reduction in DWV load in the colony. Very important for the winter bees and colony survival.

What's behind this?

There are a combination of factors and it is worth noting that the phenomena of a pre-winter clean-out was only recorded due to the use of an insert board for monitoring:-

1. The trigger is a reduction in worker brood as the queen reduces her laying rate into the autumn. Fig. 8.1b shows the frames with brood falling away from the second half of July onwards as the bees react to the declining nectar flow and pollen availability.

2. With fewer worker brood cells available, this leads to more mites on the bees rather than safely reproducing in the cells. As we saw during the broodless period following the artificial swarm, mites fall off, possibly groomed, plus there is natural mortality.

3. With fewer cells to breed in, multiple foundress mites enter the same cells to attempt reproduction. The competition leads to higher offspring mortality and reduced egg laying (Martin 1995). On the insert board, the beekeeper will see large numbers of juvenile mites (we observed up to a half of the total mite drop in some cases), in addition to the dark brick-red foundress mites. There are multiples of whole families being cleaned out. See Fig.8.4.

Fig. 8.4: The pre-winter mite reduction includes a high proportion of juvenile mites, a feature of this time of the year. Only the foundress mites are counted for monitoring purposes as they have the potential to reproduce (although many are infertile).

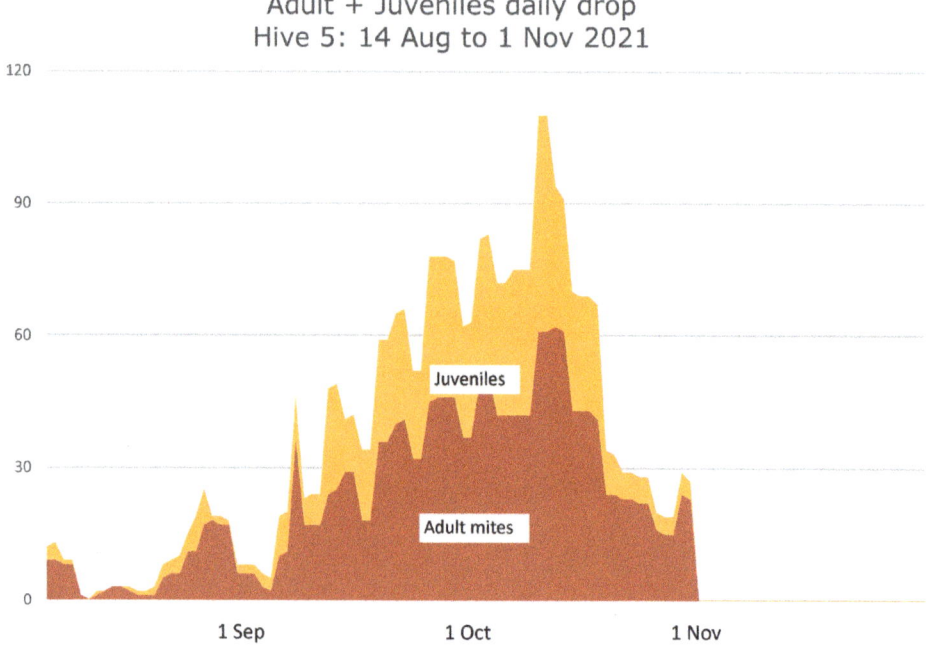

Graph: Steve Riley.

4. As well as the increased use of propolis around the brood nest at this time of year, hygienic activity against Varroa resumes following the collection of sufficient winter stores. Torben Schiffer describes this behavioural shift as 'Stores Secure' (Natural Bee Husbandry 2019). For this reason, and for nutrition, a full super of honey is left on our colonies. If too much honey is taken off, the bees will have to change focus away from hygienic behaviour.

5. Detector nurse bees have fewer brood cells to monitor, so accuracy of finding and cleaning out mites in multiple infested cells will be high.

6. Uncapping activity and chewing-out of infected pupae is intense in the August to October period. This can be seen during the end of season inspections and the monitoring of white pupae exoskeleton on insert boards. See Fig. 8.5.

Fig. 8.5: Amongst the wax clippings and scales, pollen, propolis and Varroa, lie chewed-out and discarded pupae exoskeleton targeted by hygienic bees, disrupting Varroa reproduction and ejecting infected pupae.

Picture: Steve Riley.

Winter bees

This is the critical time for winter survival of the colony. Winter bees are being raised in August to late October (in our case) and need to be healthy enough to sustain the colony through to February, March and April. The expected lifespan for winter bees is estimated at between 5 to 6.7 months (150 to 200 days: Remolina and Hughes 2008). Their key roles are to regulate the nest temperature, initiate winter brood rearing to produce the spring population and carry out early spring foraging for pollen, nectar and water.

At the start of this project, we were puzzled by how the winter bees, with such a high level of mites in August, could possibly survive to spring. Research (Dainat *et al*. 2012) on surviving winter colonies showed a rapid drop in DWV loads after reducing mites, in this research case from treatments. Extrapolating to our experience, the intense levels of hygienic activity witnessed from August onwards reduces Varroa numbers and also DWV loads. Subsequent batches of brood laid in September and October are sufficiently healthy to see the colony through to the spring. Greater tolerance of viruses, including DWV, by resistant bees may also play a role (Locke *et al.* 2021).

It would be interesting to analyse the DWV loads of our winter bees to confirm the read-through that a rapid reduction in mites leads to the same in DWV loads. One of the founder beekeeper's colonies in the project recorded a mite drop peaking at 150 per day at this time of year a few years back. The colony sailed into the following spring and had a successful season. Interesting.

Fig. 8.6: A good stock of winter bees is evident from Hive 8 on 28 January 2023, on a cold day through a perspex crown board. This is a result of the pre-winter clean out of mites by the bees in the August to October period.

Picture: Steve Riley.

<u>November-December:</u>

The last of the winter brood hatches around the 2nd to 3rd week of November. We know this from monitoring chewed-out pupae exoskeleton on the insert floor which were laid 16/17 days earlier. This is the last of the winter brood and will have suffered little from Varroa.

Mite levels are now low in a resistant colony, which is quiet going into a winter brood-break. Unlike the swarm linked brood-break, do not expect a blip up in mite drop, due to the August to October clean-out of Varroa.

In our region, new brood is laid around the 3rd week of December as the colony takes its first steps to replace itself in the spring. Over wintered mites take this opportunity, the first for about a month, to re-enter the worker brood and try to reproduce. Whilst we don't see the uncapping/recapping bees in action at this time of year, we do see evidence on the insert board of white exoskeleton of infected pupae having been chewed-out and the occasional white abdomen unceremoniously hoiked out of the front of the hive.

The beekeeper will see a gradual trending down of mite drop from natural mortality, then a drop-off, albeit from a low level, as mites re-enter brood at the end of December.

Underlying level of mites in a colony (using Fig. 8.8)

A fully resistant colony will end the year with roughly the same number of mites as it started. Levels of deformed wing virus rise and fall linked to mite levels. The bees control the mite population and DWV resulting in a host / parasite equilibrium. And a happy beekeeper!

In a susceptible colony, where there are insufficient hygienic traits against the mites, Varroa levels will be higher at the end of the year than at the start. This gives mites a higher base from which to grow the following year. As this pattern repeats itself, the bees are gradually overwhelmed, which can take 2 or 3 years. Monitoring enables the beekeeper to step in.

8 Seasonality of honey bees' hygienic behaviour against Varroa

The chart below (Fig. 8.7) illustrates the difference between bees with low or high defences to Varroa. Season on season Varroa growth in a colony increases deformed wing virus to dangerous levels. By contrast, Varroa-resistant colonies control the growth of mites and therefore DWV levels. Virus loads are measured in copies. On the right-hand scale of the chart, the rise in DWV to 10^{12} is equivalent to 1 trillion copies. DWV is elevated by mites injecting small amounts of virus during the bees' sealed brood stage which reduces the longevity of the adult bee.

Fig.8.7

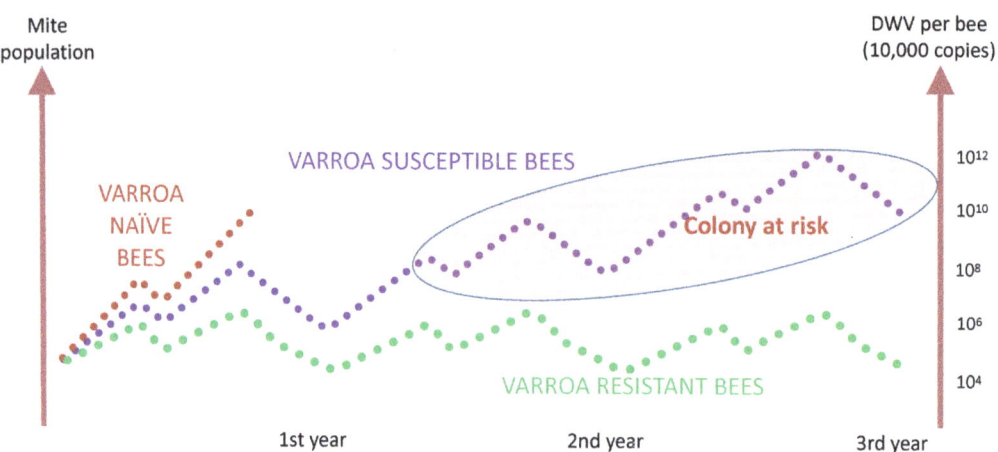

Chart: Steve Riley.

For the beekeeper, the direction of trend is important. Using the full year mite-drop data and analysis of Hive 1 in Fig. 8.1b, we have <u>estimated</u> the Varroa growth and contraction through the season shown in Fig. 8.8 overleaf.

Fig. 8.8 below shows the estimated underlying mite growth and contraction (in green) through the year for a Varroa-resistant colony. The trend in mite numbers has been estimated with reference to known mite drop data (in yellow) and correlated with worker and drone brood availability.

● ● ● *Estimated varroa growth/contraction vs mite drop*

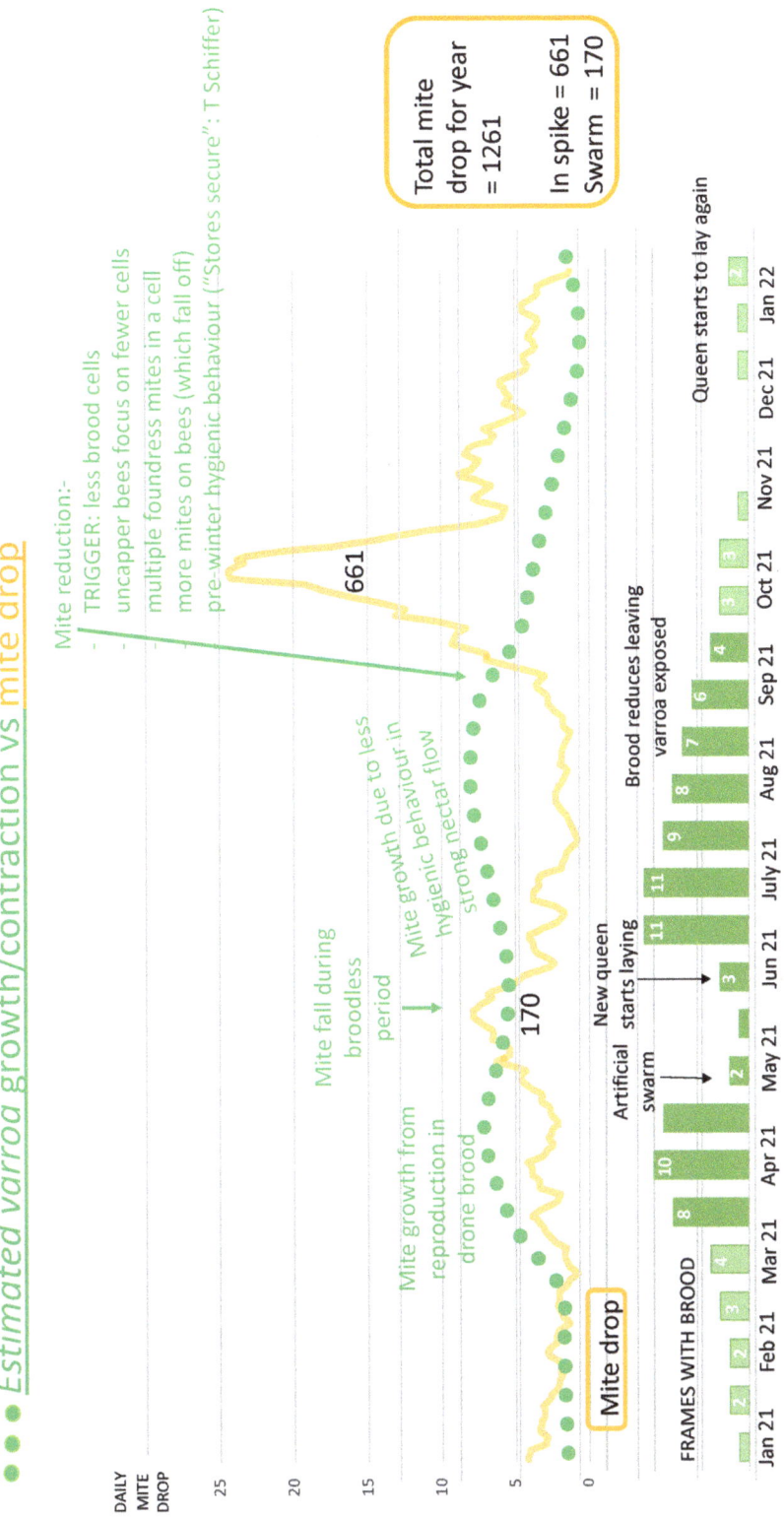

Chart: Steve Riley.

69

Interpretation of underlying mite movements through 2021:

These are the key assumptions we have made, based on the collected mite drop data, brood availability and observations of the bees' hygienic activities against Varroa:-

January-February-March

- Hygienic behaviour by bees against mites starts with the overwintered Varroa trying to breed (beekeeper observes chewed-out pupae).
- High number of infertile mites (over 40%) resulting from interrupted Varroa reproduction and high level of male mite mortality in the winter (42% vs. 18% summer).
- Resulting in a mite population failing to grow much, if at all.

April-May

- Higher mites from successful breeding during drone production where there is less or no hygienic behaviour by the bees (N.B. this relieves the worker brood to grow towards a reproductive swarm and forage on the spring pollen and nectar flow).
- Partly offset by reduced mite levels during a brood-break post the swarm; 170 mites recorded on the insert board in the peak mite drop.

June-July

- Varroa increase due to less hygienic behaviour during the strong summer nectar flow where bees prioritise winter stores and nurse bees are occupied receiving nectar, creating and storing honey.

August-September-October

- Intense bee activity against multiple Varroa families in worker brood cells, observed at final inspections (uncapping), pupae

exoskeleton and a high proportion of Varroa juveniles on the insert board.
- Pre-winter clean-out of mites; 661 of them recorded from late August to late October.

November-December

- Gradual fall away to the year end from natural mite mortality, before varroa reproduction opportunities appear at the end of December when worker brood recommences.

9
Contributory colony health factors

The Varroa-resistant traits discussed earlier in the book focus on the bees' interruption process of Varroa reproduction. There are other factors that contribute towards good colony health, most of which we use in our beekeeping. Collectively, they reduce stressors on the bees and are worthwhile, but singularly, they are not critical to achieving Varroa-resistant colonies.

Natural selection:

Honey bees left alone by beekeepers will thrive or die. They haven't needed us over their long history, although somehow, we convince ourselves that they do. *Apis mellifera*, has one of the highest genetic recombination rates recorded of any organism, aided by polyandrous mating (i.e. queens mating with lots of drones) (Beye *et al*. 2006). Adaptability is the key survival trait for honey bees.

Long-term unmanaged (or feral) colonies are therefore an important genetic resource. These bees have been through nature's sieve and the mollycoddled human-produced bees died long ago.

Rescues of colony nests from fallen trees or removals from buildings are often observed as very healthy. This was unheard of in the 1990s after Varroa arrived, when most feral colonies were wiped out. Many of today's larger collections of Varroa-resistant colonies, that are managed by beekeepers, started or benefited from wild colonies. These include the Foragers Bee & Honey Company 60-80 colonies in Worcestershire, Fen Apiaries c.100 colonies in Lincolnshire and the c.500 colonies in north-west Wales (Ref: www.varroarestant.uk case studies).

Natural selection as a strategy isn't very popular with beekeepers. But where a long-term unmanaged nest is identified, catching the swarms should include Varroa-resistant traits. Bait-hives at the ready!

Once the beekeeper has transitioned their apiary towards Varroa-resistant colonies and treatments are no longer required, there is one foot in the natural selection camp (over winter) and one foot in managed-beekeeping. In time, this should result in a tougher, more durable honey bee.

Insulation:

Most of us run British National or 14x12 wooden hives (there are a few poly hives and WBCs) at our club. These were designed to be cheaper than the double-walled hives that were popular in the late 19th and early 20th century, and lighter for migratory beekeeping. The excellent work by Derek Mitchell on the insulation properties of natural cavity nests in trees versus the heat-leaking, thin-walled wooden hives, confirms that low insulation is a stressor for the bees (Mitchell 2023). The bees have to work harder to alleviate the metabolic cost of temperature changes occurring outside of the hive that affect the internal nest. In the summer heat, this is cooling through fanning, water evaporation and bearding. In the winter when temperature stress is highest in the UK, the bees shiver their flight muscles to generate heat and also cluster. These activities to manage their internal climate use up stores for fuel, reduce longevity through increased workload and are a distraction from other activities e.g. hygienic behaviour.

9 Contributory colony health factors

Fig. 9.1: A wooden 14x12 hive belonging to Dr. T Rudd in its winter-ready format. It is cocooned with silver bubble-wrap and then 4mm correx around the sides for wind protection. An extra super above the sealed crown board houses sheep's wool and a Kingspan / Celotex square under the roof. In the complex physics equation of heat generation, insulation, condensation and ventilation, this seems to work very well. The colonies are untreated (6 years) and show Varroa-resistant traits. There has been only 1 winter loss in the apiary over that time running 4-6 colonies.

Pictures: Dr. T. Rudd.

For balance, Clive Hudson reports from north-west Wales (largest area of Varroa-resistance in the UK) that mostly wooden National and Commercial hives with solid floors are used. Little or no insulation is deployed in their mild, temperate climate.

The other interesting read-through was from the honey bees in the Arnot Forest under the observation of Tom Seeley. The bees were in tree cavities; tall, thermo-dynamic shape with good insulation. These useful properties didn't stop a large die off of colonies when Varroa arrived... Most of the bees lacked the key survival traits for Varroa.

For the record, Westerham Beekeepers tend to use insulation in the roof all year around (same principle as loft insulation) and some add cork inserts into the hand-hold sides of Nationals. The most insulation stretches to Celotex board cosies over the whole hive for winter, recognising that this time of year carries the greatest temperature stress when the colony is at its smallest. We look for inexpensive and simple insulation improvements to our (mainly) wooden hives, which we had already invested in.

Most set-ups include leaving a super of honey on for winter, where the bees naturally migrate upwards in their own warmth once brood rearing recommences in December. Honey above the bees also reduces the possibility of them losing touch with honey stored towards the cold side walls.

Brood-breaks:

Varroa need the brood of honey bees to reproduce, so brood-breaks are highly beneficial for the colony. They temporarily stop mite reproduction taking place, which lowers the future growth profile of Varroa and subsequent vectoring of deformed wing virus. Brood-breaks also stymie growth in other brood related pathogens. The opposite is also true, for those beekeepers who attempt to stop swarming.

A brood-break forces mites emerging from sealed brood cells to seek sanctuary on the bees. They are expert at hiding between the ventral abdominal plates and benefit from chemical camouflage imitating that of the honey bee. Nonetheless, Varroa are vulnerable to falling off when navigating around bees, leading to a reduction in the mite population. We always see a blip-up in mite drop during a brood-break.

This is cheering news for the beekeeper. However, most mites survive the brood-break, then reinfest when the new queen starts laying.

The number of brood-breaks across a typical season, including those from a nectar dearth or flooding of worker comb, is unlikely to have changed much over the millennia. Colonies have always benefited from brood-breaks, but that didn't help them when Varroa first arrived. Brood-breaks contribute to the health of the colony, but bees need other mechanisms to be resistant to Varroa.

9 Contributory colony health factors

Propolis:

Bees do not waste energy and resources unnecessarily. Propolis involves considerable effort, gathering resins from trees and shrubs, having it scraped off the hind legs by other bees in the nest, made into propolis and then deposited and moulded into position by the bees. In natural nests, propolis is used in large amounts at entrances and covering the inner walls, which Tom Seeley referred to as the 'propolis envelope' around brood areas in tree cavities. Research overseen by Marla Spivak (Simone-Finstrom and Spivak 2010) was able to demonstrate that the antimicrobial properties of propolis reduces the level of immune gene expression required by honey bees, produces better colony survival, higher levels of brood in the spring and improved nutritional status of young workers who feed the brood.

As beekeepers, we can rough up the insides of our brood boxes to encourage more propolis, which will contribute towards a healthier environment for honey bees. Like increased insulation, we see propolis as a positive for health, but not directly linked to why there are fewer Varroa in the colonies.

Fig. 9.2: Red propolis used around the hive entrance of a tree cavity.

Picture by Adrian Newell at the Boughton Estate Honey Bee Conservation Project.

Low intervention:

Beekeepers inspect colonies for swarming, health, space, their learning, curiosity or because they've been trained to go into the hives regularly and it fits with their work / leisure times. There is, however, a cost to the colony of the beekeeper's intervention. Bees expertly manage the temperature, humidity and CO_2 levels around the hive nest. Taking the crown board off is a stressor for the colony – a distraction from normal duties. The brood nest is split apart as frames are viewed and propolis seals broken.

Beekeeper experience reduces the number of inspections required. A lot can be learned by looking at only 1 or 2 brood frames, observing at the hive entrance or from monitoring Varroa insert boards. Does inspecting reduce the bees' ability to deal with Varroa? No. How do we know this? The beekeepers in the case studies in the section on Insulation (above) carry out inspections (not frequent) and conventional beekeeping, including taking off honey. As do we at Westerham Beekeepers. In fact, the close monitoring of the brood area has revealed much about how our bees are tackling Varroa.

Summary

There are plenty of debates in the beekeeping community on the best way to keep bees. Our guiding reference is how honey bees naturally exist, but with some compromises that come with managed beekeeping. Optimising their environment should contribute to colony health and durability. This means bees eating their own honey, surrounded by propolis, not interrupted (too much…..), in a well-insulated environment, with small defendable entrances, no open sides to their home and being adapted to the local flora, microclimate and pests.

Adding Varroa resistant traits completes the list.

10
Selection process and spreading Varroa-resistance

Selecting which colonies can manage their own Varroa populations requires judgement and the collection of behavioural and mite data. The beekeeper is building a picture of which colonies have sufficient Varroa-resistant traits or not.

The monitoring and selection process should drive rapid change, where colonies with Varroa-resistant traits take over the apiary, whilst culling (requeening) the worst.

This section focuses on our <u>ongoing</u> selection process, which will be different to when a beekeeper is getting started and transitioning away from miticide treatments, which is covered in Chapter 15.

We focus on three areas of colony evaluation:-

1) Ability of the bees to manage their own Varroa population
2) Spring brood development and foraging
3) Temperament

These are discussed below. Be careful not to select for too many criteria which can make the process bewilderingly unachievable. Keep it simple.

1) Ability of the bees to manage their own Varroa population

The bees have to overwinter without any artificial reduction of Varroa by the beekeeper, i.e. no treatments, which includes no drone culling or biotechnical measures, although you might use the latter as a short-term measure to transition off treatments. We accept (artificial) swarming as part of the bees' natural cycle and not as a manipulation to reduce mites.

Ideally, before becoming a breeder, the queen (or her daughters) would have a 3-year track record at the apiary or similar length heritage if being brought in from another apiary nearby.

In the early years or when transitioning from treated colonies, this may not be possible. Collecting swarms from long-standing unmanaged colonies can short-cut the process.

As discussed earlier in the book, observable traits for the beekeeper should include uncapping of worker brood, chewing-out of infected pupae exoskeleton on the insert board, resulting in low mite numbers. Just seeing some uncapped cells is insufficient evidence, as this is the bees' search process and mite-susceptible colonies can carry this out at a low level. Chewing-out pupae is the key to the interruption of mite reproduction and low mites are the corroboration that there is sufficient bee activity against Varroa.

The monitoring outcomes are recorded on a hive record card and integrated into everyday beekeeping. The records drive the selection process.

Fig. 10.1: Simple adaptions to the hive record card to monitor Varroa-resistant traits drives queen selection. This is free to download at the Westerham Beekeepers' website.

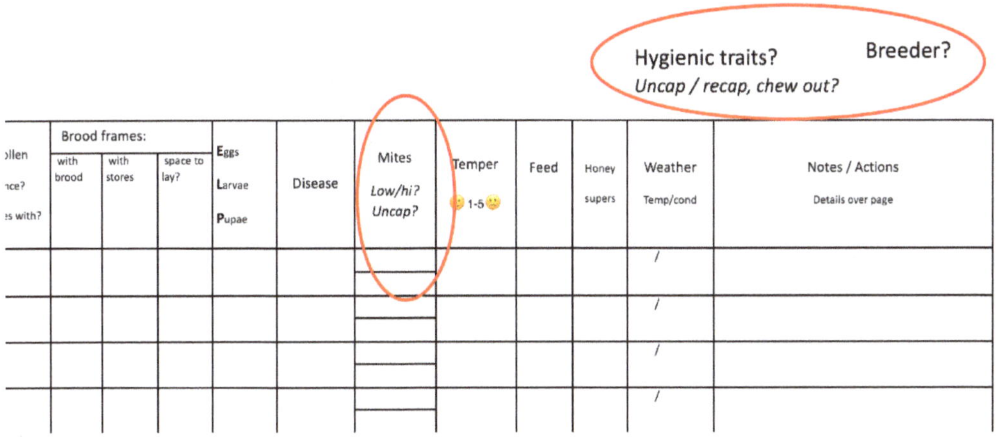

2) Spring brood development and foraging

Vigour in spring is a broad indicator of good health and demonstrates that the colony has overwintered well. An important guide in the early years, when the beekeeper is transitioning off treatments. Healthy brood should be developing strongly with lots of nectar and pollen stores coming in.

Warning signs include:-

- Slower build-up versus other similar size colonies in the apiary.
- Increasing mite numbers (e.g. 20+ drop per day).
- Signs of deformed wing virus, which can include crumpled wings, and bees crawling (or being ejected) out of the entrance and walking along the floor away from the hive.

There is a section at the end of this chapter dealing with management options for mite-susceptible colonies.

3) Good temperament

This has nothing to do with the Varroa project, but it is helpful if the colony is manageable and enjoyable for the beekeeper and not a nuisance to neighbours, users of your garden, allotment or members of the public nearby.

Our bees are marked 1 to 5. 1s are placid and quiet on the comb (and rare!). 5s are very defensive from a distance, stingy and chasers. Mostly, our colonies fall in the 2 to 3 range, with allowance made for the occasional poor behaviour due to weather or defending stores during a nectar dearth.

We see no correlation between temperament and Varroa defences, so we select bees that are good at both.

Fig. 10.2: Breeding protocol emphasises resistant traits in an apiary. Points 4 and 5 ensure other important traits are not lost.

Breeding protocol:-

1. Over-winter successfully <u>without</u> any beekeeper reduction of *Varroa*

2. Observe uncapping of worker cells <u>and</u> chewing out infected pupae

3. Low mite numbers, especially after strong brood development   Corroborates result of bees' hygienic behaviour

4. Spring health: strong brood development and nectar gathering   ) Important not to lose other

5. Gentle temperament   ) good traits

Does selecting for Varroa-resistant traits exclude other desirable traits?

This is a question we often receive. The answer is "not evidently so". Mostly, we see good all-round bees that show strong heritability in desirable traits (good foraging and brood development, Varroa-resistant and mild mannered). There is the occasional outlier colony, which is overly defensive or lacks sufficient Varroa-resistance. Both are easy to spot and can be dealt with appropriately.

Our virgin queens are from locally adapted colonies and open mated in the area. They will have 50% of their genes from her mother and 50% from a variety of local drones (which we try to influence). Together, they produce a breadth of traits and collectively provide predictability to our colonies.

I asked the same question to Clive Hudson in the north-west of Wales, where they transitioned off miticide treatments from 2008 and the answer was the same. Clive's view is that using local bees over many years has stabilised their bee population resulting in more consistent behaviours.

A loss of traits would be more of a concern for us from narrowly bred, instrumentally inseminated queens using a single drone colony – these are designed to be sold commercially and lack the diversity to be stable with open mating, in our unfortunate experience.

Heritability of resistant traits

An interesting feature of our queens has been the strong heritability of Varroa-resistant traits, i.e. they are consistently passed down the queen line from mother to daughter. This is something that we have found across all of our apiaries, many of which are unconnected to the others within the area. It was also reported by the Foragers Bee & Honey Company in Worcestershire (www.varroaresistant.uk). Good news for those raising their own queens where they are open mated (like ours).

Only a proportion of the nurse bees need to have the detector ability to identify Varroa offspring and these make up just a "tiny fraction of the bees' genome" according to Dr. J Harbo (Ref: https://www.harbobeeco.com), who led the USDA project on Varroa-resistance. An alternative view is that the detector bees have learned this ability which develops at a certain age (around 11 days old).

It is helpful to have a concentration of local beekeepers with the same selection objectives, but not that easy to achieve. We also try to influence local matings by encouraging drone production (Chapter 11).

Fig. 10.3: Co-ordinating selection methods with local beekeepers embeds resistant traits in an area.

Map data: Google, Maxar Technology.

What to do with Varroa-susceptible colonies?

Once the beekeeper starts monitoring for uncapping, chewing-out of pupae and mite loads, it will become increasingly obvious which of your colonies have low hygienic defences to Varroa. Mite levels will be rocketing and there may be signs of deformed wing virus in the hive. You do not want these bees in your apiary or in the mating area. Beekeepers are not typically very good at making tough decisions about colonies. Usually, quite the opposite. Our advice is to view them as you would a mean, defensive colony. Be proactive about effecting change towards Varroa-resistant traits taking over the apiary, whilst culling (requeening) the worst.

Here are some options for dealing with mite-susceptible colonies at different stages of the season:-

Spring (March-May):

This is the best time to take action and these are the options:

1. Requeen with your most proven hygienic stock. This will give the colony time to get on top of its mite load through the season.

    If spare queens aren't available at the time, remove or pinch out the queen in the offending colony(s) and after a week, take down any emergency queen cells raised from the old queen. Add a frame of eggs and larvae from your best colony.

2. Cull drones from the old queen to reduce their influence on local matings and avoid spreading susceptible traits.

3. If Varroa was badly out of control (e.g. mite drop 20+ per day, consider removing sealed worker brood (that will be stuffed with mites) and replace with drawn comb ideally or foundation if not available. Or treat, before requeening.

4. Move the bees out of area – this was the fate of Hive 8 in Fig. 10.4 below.

10 Selection process and spreading Varroa-resistance

Fig. 10.4: Spring monitoring will show the differences in how colonies are coping with Varroa. The graph shows the accumulative mite drop from late winter to spring with all five colonies on British National brood boxes (i.e. roughly the same size colonies). There was no winter reduction (i.e. treatment) of mites.

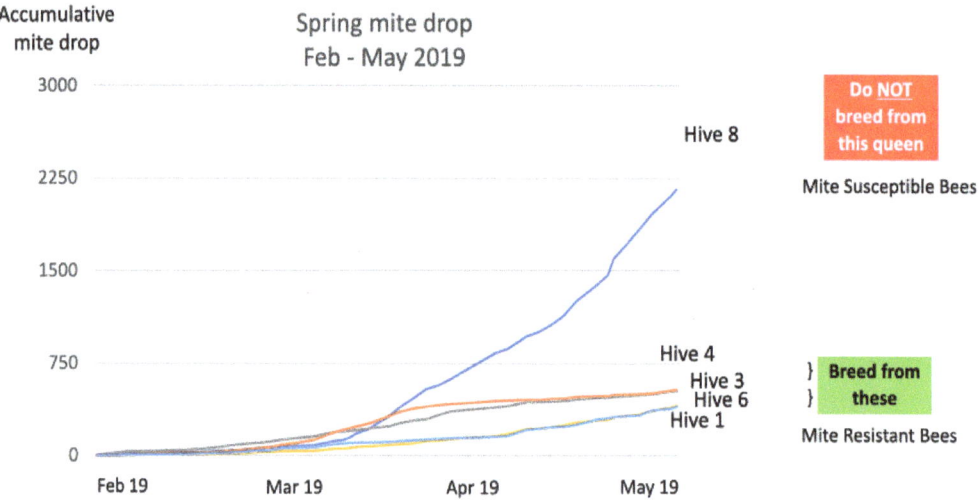

Chart: Steve Riley.

Hive 8's accumulative mite drop was over 2,200 in 3 months compared to around 500 or lower for the others. In this case, the colony was moved out of the area to a treating beekeeper. In our old regime, the beekeeper would have made efforts to keep the colony going, and if successful, would have perpetuated its weak defences to Varroa in the area.

The mite data was recorded from colonies earlier in the Westerham Beekeeper project. These days, we rarely see colonies with such high levels of spring Varroa.

Summer (June-Aug):

1. Reduce mites using biotechnical methods (e.g. queen frame trap or brood removal) or treat post honey extraction. Important to requeen next spring.

2. Pinch out queen, remove sealed brood or treat, then combine with resistant stock.

Autumn (Sep-Oct):

This may be too late to take action, as the longevity of winter bees will be compromised with a high infestation of Varroa.

1. Beekeepers' prayer (*Natural selection is my friend*). Bees often surprise us. We have had colonies with eye-watering levels of mite drop (i.e. 150 per day in autumn) in the early days of the project and they can get through. Requeen in spring unless mite levels have dramatically reduced.

2. Treat and pray; requeen in spring. This is the lower risk option and whilst it may be too late to improve the overwintering prospects, it could reduce mites being spread around the apiary in the event of the colony's demise.

11
Drones: a pivot for Varroa-resistant traits

We see drones as an important local resource to spread Varroa-resistant traits around the area. They are collectively 50% of any new mating. Breeders argue that drones are at least as important as queens and go to extraordinary lengths to produce fit and healthy ones.

In an environment where there is less or no involvement by the beekeeper to reduce Varroa, the colonies with the highest defences against mites will carry the lowest parasitic loads. They will produce fitter drones (fewer mites and DWV) and more likely be successful in drone congregation areas in passing on their genes. Drones from infested colonies have been found to weigh less and have lighter testes (Omar 2017). Fit and healthy drones are an important pivot towards spreading Varroa-resistant traits.

We seek a competitive advantage with our drones during their rearing season through optimising their comb environment to produce fit and healthy, sexually mature drones, and plenty of them. If virgin queens in the area are going to mate with 10-15 drones, we want ours to be in the mix. That means drones to be available at the start of the season for early swarms (late April) through to the main reproductive swarm period (May-June for us) and at the end of the season for late summer supersedures.

There were a few drone related issues that we sought to address:-

- Miticide impact on drone health.
- Forced-to-fit drone cells using worker foundation.
- Too few drones versus a natural level peaking at up to c.20% (Page 1981).

Miticide impact on drone health

Research has shown that miticide contamination of wax comb impacts drone:-

- Development (De Guzman *et al.* 1999)
- Survival (Rinderer *et al.* 1999)
- Sperm production (Fell and Tignor 2001)
- Sperm viability (Fisher II and Rangel 2018)

As beekeepers whose colonies do not need treatments, our concern transferred to shop bought foundation (making one's own from wax in honey supers is a good solution but quite a faff). We haven't seen an analysis of UK wax foundation, but on the assumption that most of it is recycled from beekeepers that treat with miticides, it is likely to reflect the findings of drawn comb and foundation found in US and Canadian hives, where 98% of samples were contaminated with pesticides (Mullin *et al.* 2010). The most commonly found agrochemicals in wax include beekeeper applied miticides (Payne *et al.* 2019).

Drones have the longest bee development time in the colony. From egg to adult, they develop surrounded by wax for 24 days. During this time, their sperm is produced in the larval stage through to their pupal stage (Bishop 1920), with adults emerging with all the sperm they will ever produce (Baer 2005).

Given the potential impact on drones' reproductive health, it seemed an obvious solution to allow bees to produce natural wax drone comb.

Forced-to-fit drone cells using worker foundation

Drones are more than twice the size of worker bees. Yet most drone comb building occurs using shop bought foundation designed for worker size development. This typically has a cell size of 5.4mm in diameter, whereas drone cells tend to be over 6mm up to almost 7mm. These forced cells can result in stubby and smaller drones. Smaller drones have less sperm and are less likely to mate (Gencer and Kahya 2020).

Again, the solution is in natural comb, which is right-sized for the colony's drones, or drone foundation where the wax is derived from honey supers.

Too few drones versus a natural level of up to c.20%

In managed colonies, or hives without full natural wax comb, getting the right number of drones is tricky.

What we do know is that when there is an opportunity for a colony to make additional drones, they make them. Adding an empty brood frame on the outside of the broodnest in the spring (Fig. 11.1), the bees usually make 'right-sized' drone brood comb. That's a colony signal to the beekeeper.

For just a small effort, the beekeeper can produce more healthy drones and therefore, better mated queens.

Fig. 11.1: A frame of drones. The brood frame has no foundation with the bees making the comb. The result is beautifully made right-sized cells and a healthy, pure bees' wax environment that produces thousands of drones. See Fig. 14.4 for the frame build.

Picture: Steve Riley.

At the end of the drone rearing season, the drone frame is moved further to the outside of the hive and the larger sized drone cells become efficient honey storage. This is a somewhat manufactured version of what occurs in wild colonies.

Historic malignment of drones

Much of the drones' historic malignment comes from a focus on short-term benefit strategies. One is to reduce mite loads through drone culling. Whilst this works in the season, it can contribute to poorly mated queens. A more positive approach is to deal with the underlying problem of why the mite load is high and start selecting for Varroa-resistant traits.

The other criticism levelled at drones is that they are heavy on resources. True! But from our experience, the colonies with plenty of drones are also the strongest and biggest honey producers.

High quality, healthy drones will benefit a queen's fertility and productivity. The drone contribution to the mating equation is less obvious to a beekeeper than a queen's output, but no less important.

12
Sustainable apiary for Varroa-resistant bees

We never buy bees and never expect to do so. The exception for some of us were the first beginner nucs when starting out in beekeeping. Self-sufficiency is an important element of control and influence when aiming to embed Varroa-resistant traits into the area.

What do we mean by a sustainable apiary? For us it involves:-

- Selecting and producing your own queens (keep it simple).
- Producing spare nucs to overwinter.
- Perpetually improving the quality of your honey bees – this could mean temperament, disease and Varroa-resistance.
- Effective artificial swarm control.
- Using your own locally adapted honey bees.
- Beekeepers working together when there are surpluses or shortages of bees.

In effect, you are not reliant on having to buy-in bees or take swarms from unknown sources. You control the traits in your apiary, particularly for Varroa-resistance, influence those in the locality, keep your costs low and have back-up when things don't go to plan, which we all know is a common occurrence in beekeeping.

Each main honey production colony is expected to spawn a nuc, or at least contribute to one, using either the seasonal strength in the brood or from a nuc artificial swarm.

Nuc artificial swarm

We find the simplicity of the artificial swarm method has many benefits:-

- Effective swarm control.
- A supply of spare queens.
- Keeps the foraging force together to maximise honey yield.

Removing the laying queen to a nuc stops the bees from swarming. Adding some nurse bees on a frame of sealed brood, which needs little looking after ahead of boosting bee numbers, and a frame of stores completes the move. As the nuc gains in bee numbers, foundation or natural comb can be added to encourage the nuc to build out for overwintering or other purpose. If the main colony fails to requeen successfully, there is an easy option to re-combine.

The main colony is very strong and will produce many well-nourished queen cells from its large constituency of nurse bees. Once sealed, we cut out the easy to access ones and pop them in roller cages into an incubator at 35ºC (95ºF) with a small tray of water for humidity. This little machine pays for itself very quickly. Perhaps buy one for a club or breeding group.

Fig. 12.1: Sealed queen cells cut from a brood frame and placed in the incubator. This is an old Brinsea incubator for chicken eggs.

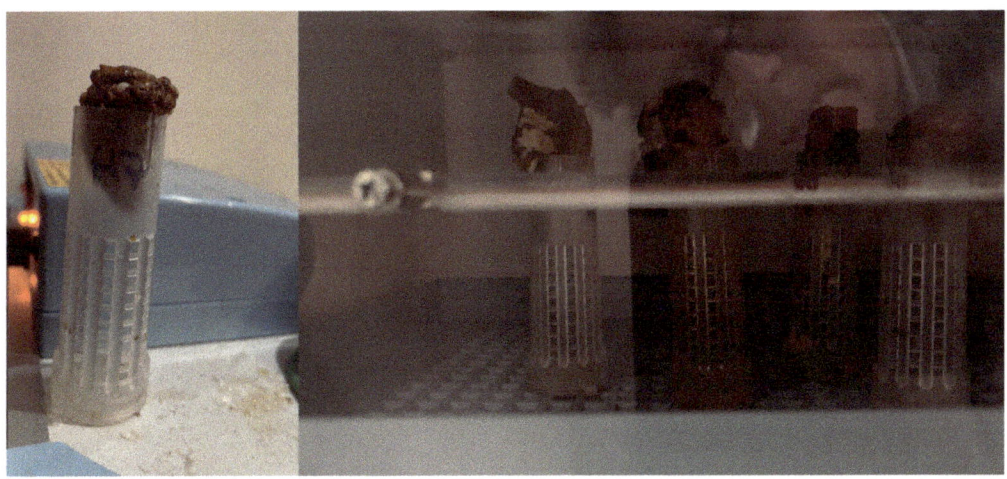

Pictures: Steve Riley.

Once emerged, the queens will feed on remnants of royal jelly inside the queen cell, but we also add honey diluted with water in the base of the roller cage.

If you don't have access to an incubator, mature queen cells (day 14+) are almost as good as virgin queens, just lacking the ease of transferability to other beekeepers. We also find the occasional dud queen cell from black queen cell virus or some other issue.

Nevertheless, beekeeper colleagues do successfully use mature queen cells to raise their queens in nucs.

We call these "Bees' choice" queen cells, a phrase borrowed from Sam Comfort of Anarchy Apiaries. Bees raising their own future queens, rather than being raised by other bees in a cell raiser hive. This author did go through a spell of grafting into a queen right colony (Ben Harden set-up) but found that too many of the subsequent queens were superseded early. This maybe a reflection of the grafting skills or perhaps bees' selection criteria of what will make a good queen has more considerations than the age and size of the larvae…

In the main colony, following a nuc artificial swarm, the foraging force remains fully intact and continues storing nectar above the brood area and pollen in the brood box.

Fig. 12.2: Using the swarm instinct and nuc artificial swarm to produce queen cells.

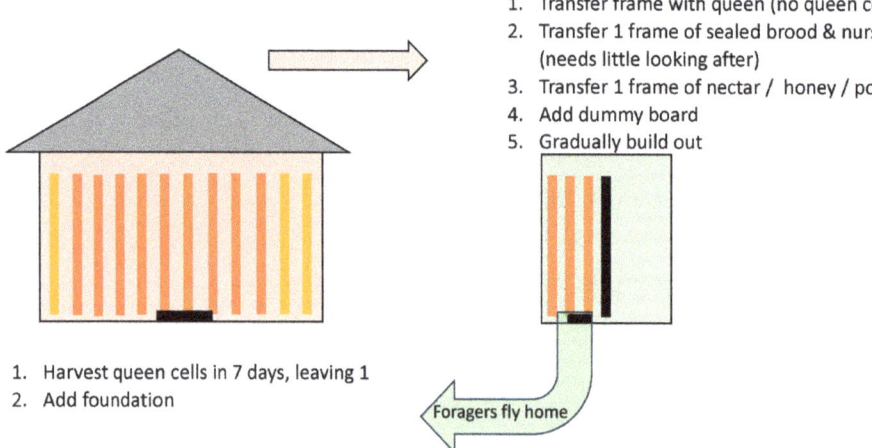

1. Transfer frame with queen (no queen cells)
2. Transfer 1 frame of sealed brood & nurse bees (needs little looking after)
3. Transfer 1 frame of nectar / honey / pollen
4. Add dummy board
5. Gradually build out

1. Harvest queen cells in 7 days, leaving 1
2. Add foundation

Foragers fly home

Chart: Steve Riley.

Historic teaching, including at our club, was to reduce (destroy) queen cells in the main colony and only leave one. It seemed a terrible waste. We now have a very simple method of producing surplus queens for our own use or sharing with club members around the local area.

Spare queens used to set up nucs

To make up additional nucs and get virgin queens mated, we use 1 or 2 frames of brood and stores from seasonally strong colonies in the apiary. These are put into nucs and dummied against the wall.

This is in preference to mini-mating nucs which, whilst using fewer bees, are a faff to set up, labour intensive and then fiddly to reintroduce into the beekeeper's set up. 1 or 2 framed nucs makes the process easier.

Virgin queen introductions:

We have found the most effective method is to lift a corner of the perspex crown board at the back of the nuc (away from any guards at the front entrance) and slip her in. Best done towards the end of the day when activity is quiet. Make up the nuc 24 hours earlier, so that the bees realise they are queenless and older forager bees have flown back to their source colony. If you want more control of the queen introduction, where perhaps the queen isn't related to the nurse bees (i.e. sourced from a different colony), put the virgin in a queen cage with the bees for 24 hours before manually releasing her once welcoming signs such as feeding are seen.

Fig. 12.3: One brood-frame nucs used to get queens mated. In the second picture, the virgin queen is being introduced in a cage hung with a cocktail stick. A frame feeder replaces a dummy board in this example.

Pictures: Steve Riley.

Full cycle sustainable apiary

Raising and selecting your own bees is very satisfying and facilitates the embedding and spreading of Varroa-resistant traits. The beekeeper keeps control of their bees, monitors and selects for their desired traits. Costs are kept low and your bees with hygienic behaviours control the Varroa.

Fig. 12.4: The full cycle sustainable apiary. It is important to be in control of Varroa-resistant traits in your apiary.

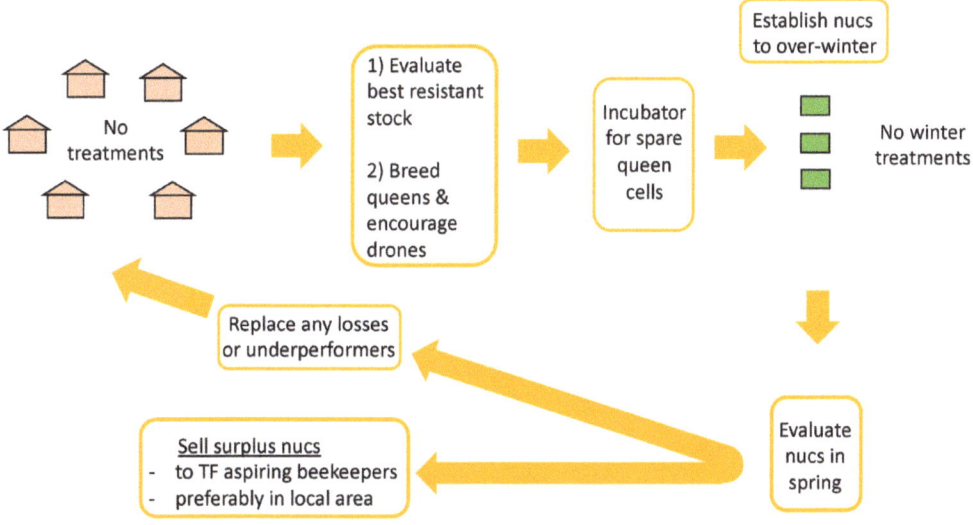

Chart: Steve Riley.

There is a simplicity to this approach that works.

We use:-

- Our own queens and their daughters, grand-daughters etc...
- Monitor for resistant traits.
- Coordinate our efforts among beekeepers.
- Use local bees adapted to the local flora and microclimate.
- Shun bought in bees, which will upset the balance already there.
- Encourage drones from resistant stocks.

Fig. 12.5: Spreading the best resistant traits across the area helps to embed the traits as widely as possible. Acting as a club, we encourage the moving of spare queens and nucs, the latter being priced at a heavy discount compared to commercial prices. We also support local clubs with Varroa projects and run apiary workshops.

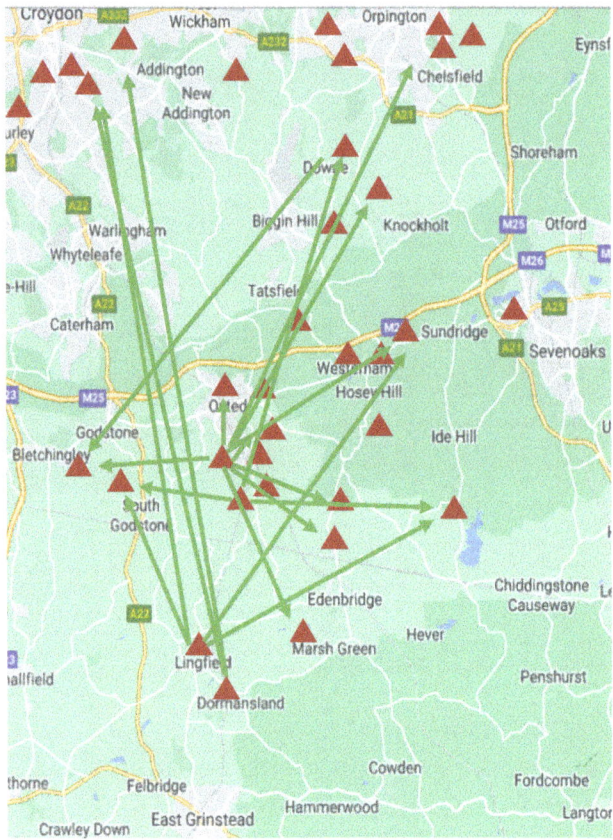

Chart: Steve Riley. Map data: Google, Maxar Technology.

13
Honey yields and colony survival

Honey yields

The apiaries in the project are dotted about the west Kent and east Surrey borders, with access to different types and amounts of forage through the season. The area can have good flows of spring to early summer nectar and pollen from local trees (blackthorn, hawthorn, horse and sweet chestnut and lime) and sometimes oil seed rape is in foraging range. A strong summer flow comes from blackberry, clover and Himalayan balsam, finishing with ivy in the autumn.

Honey production isn't a key focus for us, but a by-product of the local forage, weather and healthy bees. Our colonies are not worked hard (in a commercial sense) to maximise honey yield and there's no migration to nectar opportunities. Only locally adapted bees are used, which are naturally acclimatised to our local flora / microclimate (no imported stock). We are also increasingly aware of the need to share pollen and nectar with other pollinators.

Honey yields vary between apiaries, reflecting the colony-density to available-forage in each locality. Year on year, the main delta is the British weather and state of development of each hive through the season.

Top of the range honey yields per hive would be around 50 kg (110 lbs) at a couple of the apiaries... where everything went well, to 10kg (22 lbs) in years of poor weather and beekeeping experience (or decisions!).

These numbers exclude leaving a full super of honey for overwintering; we don't take off honey and replace it with syrup. In addition, some colonies also produce a nuc to overwinter from brood expansion earlier in the season. It is not possible for colonies to produce that level of genuine honey surplus without healthy brood.

We are sometimes asked if bees with hygienic abilities against Varroa, produce less honey. That is, are the bees spending too much time clearing out mites? We have not found that to be the case, but it's an interesting question.

The view around the founder group of beekeepers is that honey yields haven't changed noticeably since stopping miticide treatments, with seasonal weather being the main changeable factor. This makes perfect sense when thinking about what occurs inside the hive:-

- It is nurse bees that are carrying out hygienic behaviours, leaving foragers to do their work.

- The high level of hygienic behaviour was never understood or noticed by beekeepers, but had been going on for some time (see 24-hour comparison in Fig. 6.8).

- Swarm control can be achieved without reducing the immediate foraging numbers in the main colony. Using the Nuc artificial swarm method (Fig. 12.2) requires just a couple of frames of brood with nurse bees, the queen and a frame of stores (nectar and pollen). Back at the main colony brood box, nectar will be required for wax production to draw out new frames, offset by less brood to keep fed and warm until the new queen starts laying.

- The colony adjusts its behaviour and priorities. During periods of strong nectar flow, especially the summer flow which is critical for the bees' winter stores, less hygienic behaviour is observed (see Fig. 8.1b).

- Through the foraging season, a colony with resistant traits will be healthy, carrying low mite and virus loads. By contrast, Varroa susceptible colonies, in between the miticide treatments of mid-winter and late summer, suffer a rapid build-up of Varroa and deformed wing virus.

- The beekeeper has flexibility of when to take off honey – there are no miticide considerations and unsealed nectar can be left to ripen. The flexibility is also useful for holidays or when life gets in the way.

Fig. 13.1: Hives full of honey at Dr. T Rudd's apiary and Treatment free label used on the honey jars.

Pictures: Dr. T Rudd and Steve Riley.

Colony survival

Our approach to this project was to start cautiously which, with the benefit of hindsight, paid off. The gung-ho, blindly stopping miticide treatments and raising colonies from survivor stock was not for us. We had heard and continue to hear of too many examples where this approach has led to high colony losses. After 30+ years of not selecting for Varroa-resistance, the vast majority of colonies have low defences to Varroa mites. There are exceptions though, where beekeepers build out successfully from feral / unmanaged colonies or have the benefit of them nearby.

The last treatments in the colonies of the founder group at Westerham Beekeepers occurred in August and December of 2017. The first couple of winters are nervous ones. Initially, we were cautious about not putting too many of our colonies into the project and started with 28 across 8 apiaries. This grew from the second year on and averaged 40 (±5) over the period for the founder group.

Survivorship / colony losses were monitored for the founder group from 2017. The average annual survivor-loss ratio through to 2023 was 83%-17%, which compared to 77%-23% in the south-east of the UK from BBKA surveys. We assume most respondents in the area surveys do treat. It surprised us that our colony survivorship was that good overall and better than the average of local performances.

There were a range of experiences within our group. One apiary recorded only 1 colony loss over the period, whilst another, where there had been notably less selection for traits, has struggled. Another apiary had early high losses, then introduced honey bees from some dilapidated hives which had been unmanaged for some years on a nearby aerodrome. What a difference they made!

Where there are colony losses, working as a group pays off. Queens and nucs with resistant traits are being transferred to the apiaries most needing them.

Importantly, we now have a strong breeding base from which to spread resistant traits in our area, where colonies have increased to around 100 involved in the project. Beekeeper expertise to identify Varroa-resistant traits is also growing through education, knowledge share and experience. Our strategy is to build out more of the bees with the right traits and increase our queen raising capacity in 2024 to make more queens and nucs available locally.

Colony outcomes from other beekeepers who have had a longer-term experience of not treating show a similar or better overwintering out-performance versus their local peers.

In north-west Wales, at the Lleyn & Eifionydd BKA, Clive & Shân Hudson ("thank you" for this huge admin task) monitored 1,573 treated colonies and 1,096 non-treated colonies over the 5 years from 2010 to 2015. The average annual loss rate for non-treated colonies was 13% versus 19% for treated. This is from the largest area of Varroa-resistant bees in the UK, where colonies number around 500 (https://beemonitor.org/).

Similar overwintering performance versus the local beekeepers was reported by the Foragers Bee & Honey Company in Worcestershire. 2007 was their first year without Varroacides and biotechnical methods were used to transition to treatment-free. Winter losses of full colonies ranged from 7% to 12% in 2019/20 to 2021/22 stating that: "Winter losses were usually the same as those reported by other local beekeepers or sometimes lower." (Ref: www.varroaresistant.uk and BBKA Spring Convention presentation April 2023).

The bees in Varroa-resistant colonies are not invincible but are tested over winter without mite reduction from the beekeeper. This element of natural selection, together with tolerance to Varroa vectored viruses (Locke *et al.* 2021), should create a tougher honey bee in time. Colony survivorship records are starting to bear this out.

As a guide to what is possible, in South Africa, where Varroa arrived in 1997 and nobody treated, annual losses subsequently reverted back to the pre-Varroa days at around 5% per annum (Allsop 2006). There were upfront losses of about 30%, which is what our conservative starting strategy sought to avoid.

14
Lessons learned

There was no educational pathway to follow when we started this project, so here are a few of our wrong turns, which hopefully will short-cut your own learnings.

Surviving 3 winters

3 winters without miticide treatment or beekeeper mite reduction is a strong indication of Varroa-resistance, especially if mite numbers remain low in the third year and the key traits are recognised. We were cock-a-hoop with excitement after the first year of survival, had calmed down after the second year and learnt (the hard way) which were our best bees after the third. It is important to see the key traits of resistance all the way through.

Be tough and decisive with culling underperforming queens

Don't fall in love with your favourite colonies! You do not want their traits (or drones) out there facilitating large build-ups of Varroa and deformed wing virus. Act in the same way as you would towards a mean and defensive colony. Requeen in the first half of the season to allow the new queen's progeny to takeover ahead of winter preparations.

Colonies with high mite loads carry a potential jeopardy to the rest of the hives in the apiary, particularly over winter. For example, if bees from a collapsing colony abscond on a mild January day to a hive with very high defences to Varroa, the bees in the receiving hive will uncap and chew-out pupae into the spring and are unlikely to develop or will collapse themselves.

Seeing uncapped worker brood isn't enough

Even some Varroa-susceptible colonies will uncap. Seeing uncapping during inspections is insufficient evidence to stop treating. Look for chewed-out pupae and corroborate the sufficiency of the colony's hygienic traits against Varroa through low mite numbers.

Grooming and mite mutilation - important?

Most research on Varroa-resistant bees concluded that grooming and mite mutilation are important traits. To us beekeepers this sounded logical and efficient. The bees would groom and kill the mites before they had a chance to reproduce, chewing off legs or other body parts. We would all cheer that! At Purdue University in the USA, they bred for the trait and their bees were known as "Ankle-biters" (McAfee 2021).

As a consequence, we spent large amounts of time peering down low power microscopes at mites found on the bottom board. Investigations revealed what we thought was evidence of grooming and mutilation. There was one hive where between a quarter and one third of the mites were mutilated. Legs or antennae were missing, or the carapace was obviously damaged.

Fig. 14.1: Mites damaged by the bees at Westerham Beekeepers.

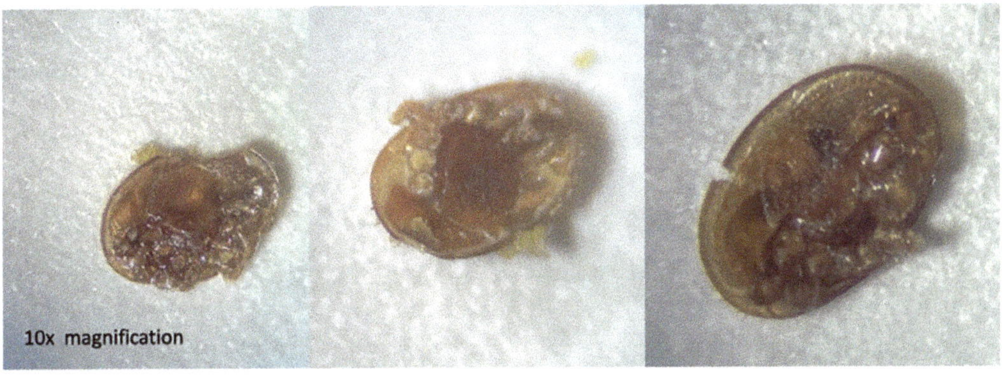

Pictures: Steve Riley.

14 Lessons learned

The Varroa pictured above were from the author's Hive 5. This colony was headed by a beautiful large queen, produced lots of brood, large amounts of surplus honey and the bees were gentle to handle. We even bred from her. After surviving 2 winters and 3 seasons without treatments, it felt like this was the real deal – mite grooming and mutilation in practice.

What we couldn't understand was why there were large amounts of mites in the colony – in fact, the highest numbers in the apiary by some margin. Something didn't add up.

Fig. 14.2: There seemed no correlation between observed mite mutilation and low mite numbers.

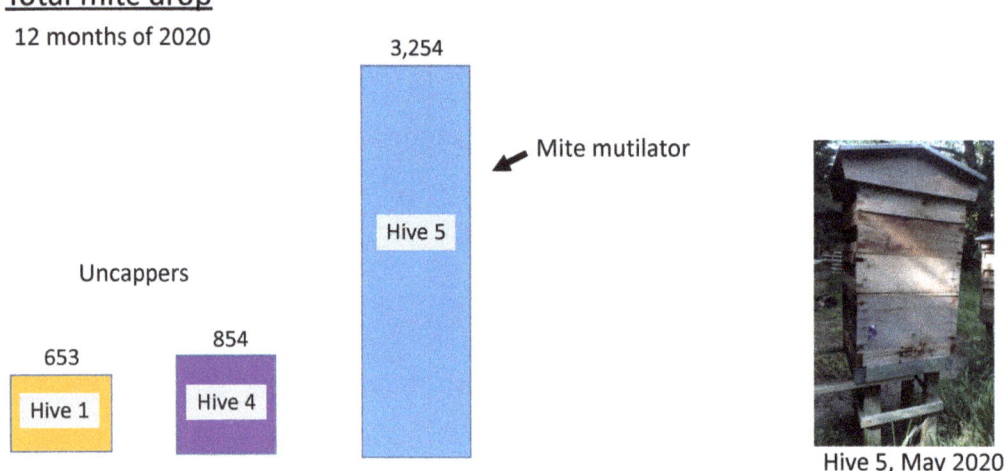

Hive 5, May 2020

Graph and picture: Steve Riley.

The answer became apparent in the June of the queen's 3rd season. Supersedure cells. Her time was up. She was retired to overwinter in a nuc to see if her genetics could be retained for the following season. The colony requeened and the daughter was laying beautifully. Stores were sufficient and all was set for the winter... The colony was dead by Christmas. One of those awful moments in beekeeping. Reality had met optimism. What had happened?

There was further evidence to consider:-

- 3 nucs had been produced led by daughter queens from Hive 5 and gone to other apiaries. None of them overwintered successfully.
- Uncapping / recapping traits were recorded during inspections, as well as mite mutilation, but <u>little chewing-out of pupae was observed</u>.
- The ageing, original queen successfully overwintered in the nuc!

Hive 5 post mortem conclusions:-

1. Mite mutilation / grooming is <u>not</u> a key contributor to bees reducing Varroa loads.

2. Damaged Varroa are more likely a result of bees cleaning out infested cells, particularly softer body immature mites. These bees had a lot of cleaning out to do.

3. The level of uncapping / recapping and chewing-out of pupae was insufficient in that colony to control mite numbers.

4. Tolerance to viruses was possible, which probably explained why the colony survived with a high mite load and why the ageing queen outlived her daughters.

5. Colonies with high levels of mites carry the deformed wing virus jeopardy and are a risk to the apiary. Requeen.

Subsequent research has cast doubt on the role of grooming in Varroa-resistance (Grindrod and Martin 2023). Much of the assumption that grooming was an effective method of reducing Varroa came from small sample studies of *Apis cerane*, the Asian honey bee. Other research found little difference between the level of grooming between susceptible or resistant *Apis mellifera* bees (Aumeler 2001).

Varroa have a clever ability to chemically camouflage themselves by mimicking the bees' cuticular hydrocarbon profile (smell) down to the level of each colony (Kather *et al.* 2015). They are quick, agile, and have a flattish button shape which enables them to squeeze between the ventral abdominal plates of a honey bee to feed and also escape from grooming.

Fig. 14.3: Varroa between the ventral abdominal plates of a honey bee.

Picture: Courtesy The Animal and Plant Health Agency (APHA), Crown Copyright.

Grooming may still have a minor role in Varroa-resistance as bees are known to groom themselves and each other. It is not something that we select for and if mite levels are high in a colony, then we assume that other hygienic behaviours against Varroa are insufficient. These colonies are requeened.

Small cell:

This is an interesting one that divides the class. What sparked our initial interest in small cell was an article by beekeeper John White, in the June 2019 BBKA Magazine, on the use of small cell foundation. He had not used miticide treatments for 9 years and attributed the Varroa-resistance to smaller or more appropriate size honey bees resulting from smaller cell foundation.

Elsewhere, there seemed, on the face of it, a link between smaller bees and Varroa-resistance. In Asia, where Varroa and *Apis ceranae,* live in equilibrium, the natural cell size is 4.2 - 4.8mm. Varroa-resistant Africanised bees in the USA come in at 4.4 – 4.9mm. The bees that survived Varroa in the Arnot Forest were described as "markedly smaller" (Seeley 2019). This particularly interested us as the bees were exposed to natural selection. Back in the UK, Joe Ibbertson at the Boughton Estate Honey Bee Conservation Project had also observed very small honey bees living in tree cavities.

Claimed benefits of small cell included:-

- Longer living bees (more bees in brood area doing less work).
- 1 day shorter development time (i.e. 20 versus 21 days) leaving less time for Varroa offspring to mature and be viable.
- Damage to the only male Varroa in the smaller cell, as it moves to the feeding area. This would stop reproduction taking place.

We couldn't resist and decided to investigate small cell further in our own hives.

After 3 seasons and 3 winters on small cell foundation throughout 2 Westerham Beekeeper apiaries, our observations were:-

1. We observed no material change in the level of uncapping / recapping which continued to vary by colony.

2. Varroa was lower over the last 3 years across our apiaries, which was also true in our apiaries using large cell.

3. Smaller bees in surviving feral / unmanaged colonies (e.g. Arnot forest, Blenheim Estate) were likely a result of old comb producing smaller cell diameters from the build-up of propolis and old larval moults, rather than a conferred natural selection advantage to dealing with Varroa.

Our conclusion was that the key traits of uncapping plus chewing-out pupae were not bee-size dependent. Emeritus Professor Stephen Martin, in his research capacity, has observed Varroa-resistant *Apis*

14 Lessons learned

mellifera bees in different parts of the world and confirms that resistant bees come in "*different sizes and colours*".

Using *wrong size*, shop bought foundation is an issue easily overcome by allowing bees to build their own comb in the brood area – increasingly, this is what we are doing.

Fig. 14.4: A brood frame without foundation allows bees to make the right-size worker or drone comb. Lolly sticks and a 4mm BBQ skewer provide guides and stability. A little wax is rubbed around the inside of the frame to encourage the bees.

Picture: Steve Riley.

15
Getting started / transitioning from treated colonies

Once the decision was made to stop miticide treatments, we felt energised to find the solution, which we knew had occurred in nature. Observing bee behaviours with Varroa was educational. It was a return to traditional beekeeping, replacing the know-how and equipment of applying miticides to colonies. It is also the long-term solution to Varroa.

In the early years of stopping treatments, we found mite counts were higher than comfortable in some colonies, particularly in the autumn on the critical winter bees. It took us a few years to understand the normality of the pre-winter spike in cleaning out Varroa, as research had never covered this previously. Selecting for Varroa-resistant traits seen in naturally resistant bees was a new area of beekeeping and with hindsight, we took more risk with our colonies than we are recommending here.

By way of example, Fig. 15.1 shows the mite drop in Hive 4 over the first two years of no Varroa reduction by the beekeeper. The Varroa load was high in the first full year, averaging 9 per day (we like to see 5 or under now). Encouragingly though, we were seeing uncapping during inspections and knew from earlier research on Varroa sensitive hygiene by the USDA that this was an important trait (Harbo and Harris 2005).

At the start of the second year, chewed-out pupal bits were appearing on the insert board, which we had started to understand was an important part of the bees' interruption process of the mites' reproduction.

The second year showed a 75% drop in mite levels which is over 2,500 fewer Varroa. The bees had reacted to their mite problem – they had capabilities that they had not previously needed to show, and we had not realised were there. Was this an example of epigenetics, where traits are turned on as required? We were pleased to have their traits in the apiary and their drones flying. Our judgement had benefited from observations from the earlier research, holding our nerve and a slice of good fortune.

Fig. 15.1: The first two years of mite drop without any Varroa reduction by the beekeeper.

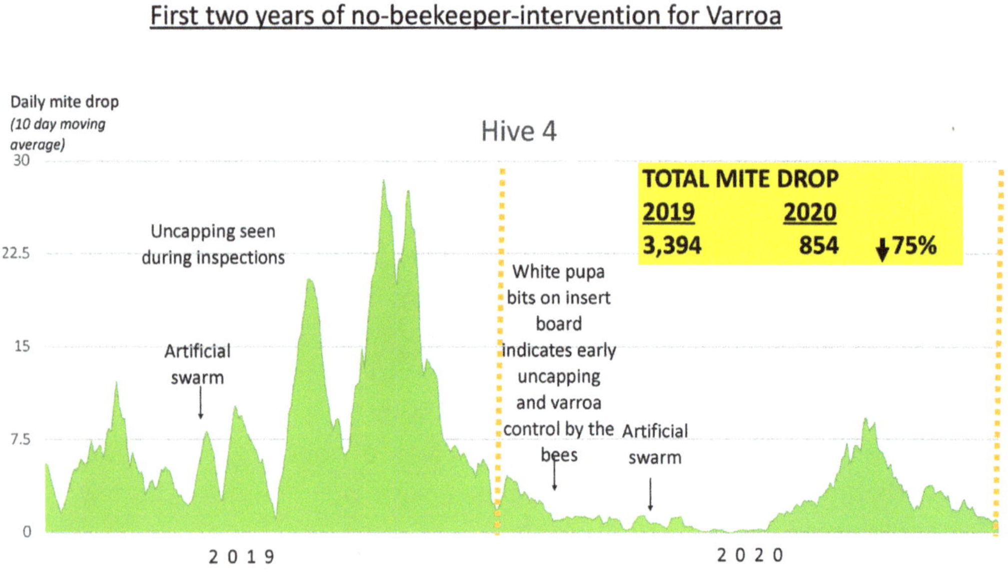

First two years of no-beekeeper-intervention for Varroa

Hive 4

TOTAL MITE DROP
2019	2020
3,394	854 ↓75%

Chart: Steve Riley.

Varroa levels declined over the project period

We are now in the seventh year of the Varroa project. Our selection and observational skills have improved over that period and as a result, Varroa levels are generally lower.

If you are new to monitoring and selecting for Varroa-resistant traits in your colonies, expect to make a few mistakes along the way. We did. Judgement errors are easier to see in hindsight. In Fig. 15.2, we were too fascinated by the possibility of mite mutilation (wrongly) in Hive 5. There were two ignored clues to their low defences to Varroa; firstly, from the higher levels of mite drop at the start of the year and high previous year Varroa levels. Secondly, from seeing less uncapping and chewed-out pupae.

Fig. 15.2: The chart shows the mite drop data for different hives in the same apiary in 2020, 2021 and 2022. Generally mite levels are lower due to selection and proliferation of the best colonies with de-selection and loss of the worst. Improved judgement comes with monitoring and experience.

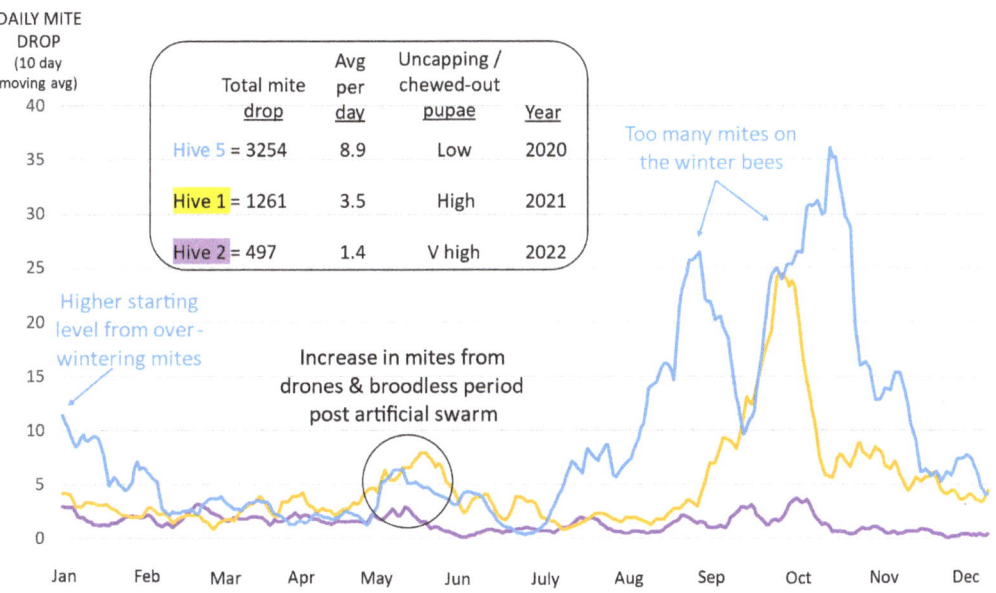

Chart: Steve Riley.

Transitioning from miticide treatments to Varroa-resistance

If you have got this far through the book, the mechanisms the bees are deploying to interrupt Varroa reproduction will be known, as will the observations required by the beekeeper to interpret this interaction.

Next follows a step-by-step plan, including learnings from the Westerham Beekeeper project, of how to proceed towards changing the traits in your apiary or at your club. Do adapt them to suit your own style of beekeeping.

1. Be cautious. For over 30 years, there has been little selection for Varroa-resistance (bar unmanaged colonies) resulting in a large majority of honey bees with low defences (i.e. Varroa breeders). This includes most commercially sold bees, so do ask about the

treatment regime and tolerance to Varroa before you buy (N.B. better to raise your own).

The seven founder members at Westerham Beekeepers started by putting half of their hives into the project, which felt like a sensible approach at the time and still does.

We find many of our newer beekeepers are keenest to go down this route. Some make it through with good mentoring, but generally it is best to learn the bee-craft first. The early years have enough hazards!

2. Be prepared for a nervous first few winters. Changing beekeeping practice magnifies the normal anxiety over colony loss that beekeepers suffer at this time of year.

3. Consider transitioning off miticide treatments using biotechnical methods (e.g. queen frame trapping replacing the summer treatment) which overcomes the early years' fear that *if you don't treat, your bees will die* (full details in the Appendix).

    Move on quickly from biotechnical Varroa reduction as it becomes a miticide substitute, masking which colonies have developed the defences to cope with Varroa or not.

4. Monitor for uncapping, chewed-out pupae = low mites (minimum of monthly). Wait until you have clearly identified Varroa-resistant traits in your bees before stopping all beekeeper mite reduction. Spend a season monitoring if necessary.

5. Choose a convenient apiary that's easily accessible to start the monitoring process. A garden apiary at home is ideal. One neighbouring club set up a new apiary to start their project.

6. Start monitoring a minimum of 4-5 colonies. Some will demonstrate more hygienic Varroa abilities than others with their new regime. Group into categories of HIGH (defence against Varroa), LOW or UNDETERMINED. The HIGH colonies will become the breeding stock to take over the apiary genetics in future seasons.

7. Integrate monitoring into day-to-day beekeeping through headings on a hive record card. Include your assessment of the queen as a breeder and record the progress of her daughters.

8. Stop the winter treatment initially on selected colonies where you are already seeing uncapping and chewed-out pupae. The spring is a revealing time to be able to judge how your bees are coping with Varroa after 6-8 months together, following any mite reduction by the beekeeper the previous summer.

9. Raise your own queens to control Varroa-resistant traits in and around the area of your apiary. An incubator is very helpful to hatch spare queens from chosen colonies and pays for itself very quickly.

10. Raise plenty of drones from your best colonies with Varroa-resistant traits to embed them locally. Use natural comb to avoid miticide in foundation and to right- size the cell. Cull drones from colonies with low Varroa defences to avoid spreading their traits.

11. Produce spare nucs to over winter from colonies with the most resistant traits. These are good back-ups and help the apiary become self-sustaining.

12. Increase from your best survivor stocks in the spring (evident resistant traits, low mite counts, strong brood and foraging) to replace any winter losses. Monitoring will make the proliferation of daughter queens from your best colonies very obvious, a good example being from Hive 2 in Fig. 15.3 below, where her family are now the dominant queen-line in the apiary.

13. Actively manage colonies that have low Varroa defences and high mite loads, especially in the late summer where they can create a winter jeopardy from mites leaving from a collapsing colony. Short-term measures to reduce Varroa could include biotechnical (e.g. queen frame trap or brood removal) or to treat. Then, requeen in the spring.

15 Getting started / transitioning from treated colonies

14. Use locally adapted bees suited to your area's flora and climate. The adaption we are seeing in our colonies will also be taking place in others around the country and elsewhere, where bees are being allowed. This means where there is the continuity of using locally adapted bees from your area. Buying in shop bought bees, which are often based on an instrumentally inseminated imported breeder queen (but badged British), loses that continuity.

    In the case studies of Varroa-resistant bees in the UK (www.varroaresistant.uk), **every example** uses local bees from their area.

15. Work as a group sharing the best ideas and queens. Commit to helping beekeepers with losses, which also spreads the best traits. A number of beekeeping clubs have already started their own projects, including setting up specific apiaries.

16. Source bees from long-standing feral / unmanaged colonies that have been exposed to natural selection. Use bait-hives nearby or be first to any swarms! This is a short-cut to Varroa-resistant bees.

17. Be cautious of introducing swarms from unknown origins; they are likely to be bees with low mite defences. If you do house a swarm, check for resistant traits and requeen if necessary.

18. Be particularly friendly to beekeepers who already have Varroa-resistant bees – for queens, nucs or advice.

19. Overwintered nucs are a good platform for getting started from beekeepers with a long-term track record of not treating (over 5 years ideally):-

    a) The queens are mated at the source's apiary (where there are resistant traits).
    b) There could be an opportunity to make an extra colony and possibly also a nuc during the first season to overwinter.
    c) Drones will be produced, starting to influence the local matings in the area of your apiary.

20. Whilst not directly linked to the key Varroa-resistant traits, encouraging propolis, adding insulation and low intervention will reduce stressors on the colony. This optimises the bees' environment, contributing to their health and survivorship.

21. Take a 3-year view on the project to change your apiary traits and 5-year view to embed them locally. The beekeeper can quickly change the apiary dynamics by promoting colonies with resistant traits and requeening those without. Embedding the traits locally takes longer as you are seeking to influence the local mating areas with your drones. Call on your beekeeping neighbours to get them involved!

We will finish by demonstrating what selecting for Varroa-resistant traits can achieve, using relatively small numbers of locally adapted colonies (in this case, around 10 colonies in one apiary).

Fig. 15.3: Two years of monitoring over 2022 & 2023 shows mite levels so low in Hive 2 that Varroa can only be having limited success at reproducing in worker brood.

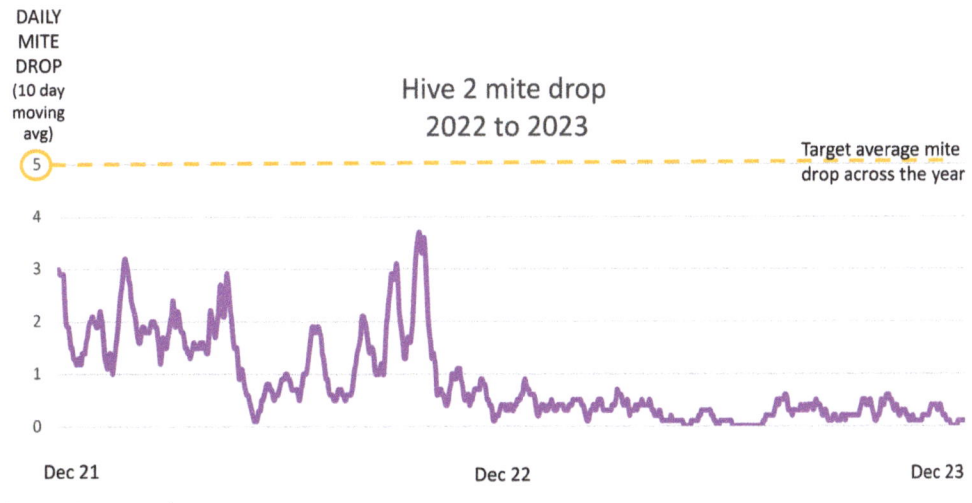

Chart: Steve Riley.

This is evidence that *Apis mellifera*, the European honey bee, can produce a similar host / parasite equilibrium seen in the original host, *Apis cerana* in Asia.

15 Getting started / transitioning from treated colonies

Good luck with learning about the capabilities of your honey bees! Enjoy making a difference and contributing to *Apis mellifera* solving its Varroa problem.

Acknowledgements

As hobbyist beekeepers, we were fortunate to have Dr. Ralph Büchler as our mentor for the start-up of the project. Ralph was Director of Research, with a focus on bee breeding, at the Kirchhain Bee Institute in Germany. He brought decades of research experience to Varroa issues and always responded to our emails – thank you.

Emeritus Professor Stephen Martin initiated and led the "Path to Varroa-resistance in the UK" project following the launch in April 2023 of a free-to-use website: www.varroaresistant.uk. He lectures and presents on Varroa-resistance around the UK and internationally. Stephen helped us home in on and understand the key resistant traits. It's been an honour to share the occasional platform with him and thank you for editing the science sections of this book.

Thank you to Rhona Toft for the diligent editorial input, Joe Ibbertson and Clive Hudson for the support, encouragement and for so generously sharing your expertise.

The founder team at Westerham Beekeepers came together in the Autumn of 2017 at the Carpenters Arms: Alban, Annie, Jacky, Keith, Mark and Topsy. We hiked up some stiff learning curves together, enjoying the highs but importantly, sharing the lows.

Thank you to my wife, Christine, for her loving support and editorial input. Sharing a house with someone writing a book is not easy!

And finally….. the honey bees, that embellish all of our lives as beekeepers and who have communally found a solution to Varroa. All we did was to follow their lead.

Acknowledgements

Fig. 16.1: The queen and daughters from Hive 2, the most Varroa-resistant bees we have seen so far.

Picture: Steve Riley.

References and further reading

Allsopp, Mike. Analysis of Varroa destructor infestation of southern African honey bee populations. University of Pretoria (South Africa), 2006.

Anderson, D., Trueman, J. Varroa jacobsoni (Acari: Varroidae) is more than one species. Exp Appl Acarol 24, 165–189 (2000). https://doi.org/10.1023/A:1006456720416

Aumeler Pia, Bioassay for grooming effectiveness towards Varroa destructor mites in Africanized and Carniolan honey bees, Apidologie, 32 1 (2001) 81-90, DOI: https://doi.org/10.1051/apido:2001113

Baer, B. (2005) Sexual selection in Apis bees. Apidologie 36, 187–200.

Bailey L, Ball B V, VIRUSES, Honey Bee Pathology (Second Edition), Academic Press, 1991, Pages 10-34, ISBN 9780120734818, https://doi.org/10.1016/B978-0-12-073481-8.50007-2.

Beye M, Gattermeier I, Hasselmann M, Gempe T, Schioett M, Baines JF, Schlipalius D, Mougel F, Emore C, Rueppell O, Sirviö A, Guzmán-Novoa E, Hunt G, Solignac M, Page RE Jr. Exceptionally high levels of recombination across the honey bee genome. Genome Res. 2006 Nov;16(11):1339-44. doi: 10.1101/gr.5680406. Epub 2006 Oct 25. PMID: 17065604; PMCID: PMC1626635.

Bishop G H, (1920) Fertilization in The Honey-Bee I. The male sexual organs: their histological structure and physiological functioning. J. Exp. Zool. 31, 225–264.

Crailsheim-Karl, Robert Brodschneider, Pierrick Aupinel, Dieter Behrens, Elke Genersch, Jutta Vollmann & Ulrike Riessberger-Gallé

(2013) Standard methods for artificial rearing of Apis mellifera larvae, Journal of Apicultural Research, 52:1, 1-16, DOI: 10.3896/IBRA.1.52.1.05

Dainat B, Evans JD, Chen YP, Gauthier L, Neumann P. Dead or alive: deformed wing virus and Varroa destructor reduce the life span of winter honeybees. Appl Environ Microbiol. 2012 Feb;78(4):981-7. doi: 10.1128/AEM.06537-11. Epub 2011 Dec 16. PMID: 22179240; PMCID: PMC3273028.

De Guzman L I, Rinderer T E, Lancaster V A, Delatte G T, Stelzer A, (1999) Varroa in the mating yard: III. The effects of formic acid gel formulation on drone production. Am. Bee J. 139, 304–307.

Donzé, G., Guerin, P.M. Behavioral attributes and parental care of Varroa mites parasitizing honeybee brood. Behav Ecol Sociobiol 34, 305–319 (1994). https://doi.org/10.1007/BF00197001.

References and further reading

Donzé, G., Herrmann M., Bachofen B., and Guerin P. M.. . 1996. Effect of mating frequency and brood cell infestation rate on the reproductive success of the honeybee parasite Varroa jacobsoni. Physiol. Entomol. 21: 17–26.

Ellis M B, Nicolson S W, Crewe R M et al. Brood comb as a humidity buffer in honeybee nests. Naturwissenschaften 97, 429–433 (2010). https://doi.org/10.1007/s00114-010-0655-1

Evans D, Apiarist Blog, (2019), Mites equal viruses.

Fell R D, Tignor K, (2001) Miticide effects on the reproductive physiology of queens and drones. Am. Bee J. 141, 888–889.

Fisher II, A, Rangel J, (2018) Exposure to pesticides during development negatively affects honey bee (Apis mellifera) drone sperm viability. PLoS ONE 13, e0208630.

Fuchs, S: Preference for drone brood cells by Varroa jacobsoni Oud in colonies of Apis mellifera carnica, Apidologie, 21 3 (1990) 193-199,DOI: https://doi.org/10.1051/apido:19900304

Gençer H V, Kahya Y, Sperm competition in honey bees (Apis mellifera L.): the role of body size dimorphism in drones. Apidologie, 2020, 51 (1), pp.1-17. ff10.1007/s13592-019-00699-4ff. ffhal-03006580f

Grindrod I and Martin S J. 2021 Parallel evolution of Varroa-resistance in honey bees: a common mechanism across continents? Proc. R. Soc. B 288: 20211375. https://doi.org/10.1098/rspb.2021.1375

Gilliam M, Taber III S, Richardson GV, Hygienic behaviour of Honey Bees in relation to Chalkbrood disease, Apidologie 14 (1983) 29-39, https://doi.org/10.1051/apido:19830103

Gramacho K, Spivak M (2003), Differences in olfactory sensitivity and behavioral responses among honey bees bred for hygienic behavior, DO - 10.1007/s00265-003-0643-y, Behavioral Ecology and Sociobiology

Grindrod I, Martin SJ, (2021) Spatial distribution of recapping behaviour indicates clustering around Varroa infested cells, Journal of Apicultural Research, 60:5, 707-716, DOI: 10.1080/00218839.2021.1890419, https://doi.org/10.1080/00218839.2021.1890419

Grindrod I and Martin S J. 2021 Parallel evolution of Varroa-resistance in honey bees: a common mechanism across continents? Proc. R. Soc. B 288: 20211375. https://doi.org/10.1098/rspb.2021.1375

Grindrod, I, Martin, S J, (2023). Varroa-resistance in Apis cerana: a review. Apidologie, 54, https://doi.org/10.1007/s13592-022-00977-8

Harbo Bee Co.: Breeding resistance to Varroa mites https://www.harbobeeco.com/

John R Harbo & Jeffrey W Harris (2005) Suppressed mite reproduction explained by the behaviour of adult bees, Journal of Apicultural Research, 44:1, 21-23, DOI: 10.1080/00218839.2005.11101141

Hawkins G P, Martin S J, Elevated recapping behaviour and reduced Varroa destructor reproduction in natural Varroa-resistant Apis mellifera honey bees from the UK. Apidologie 52, 647–657 (2021).

Hudson C, Hudson S. Treatment-free beekeeping. British Beekeepers Association News 2020; 277: 229–232.

Human H, Brodschneider R, Dietemann V, Dively G, Ellis JD, Forsgren E, Fries I, Hatjina F, Fu-Liang Hu, Jaffé R, Bruun Jensen A, Köhler A, Magyar JP, Özkýrým A, Pirk CWW, Rose R, Strauss U, Tanner G, Tarpy DR, van der Steen JJM, Vaudo A, Vejsnæs F, Wilde J, Williams GR & Huo-Qing Zheng (2013) Miscellaneous standard methods for Apis mellifera research, Journal of Apicultural Research, 52:4, 1-53, DOI: 10.3896/IBRA.1.52.4.10

Kather R, Drijfhout FP, Shemilt S, Martin SJ. Evidence for passive chemical camouflage in the parasitic mite Varroa destructor. J Chem Ecol. 2015 Feb;41(2):178-86. doi: 10.1007/s10886-015-0548-z. Epub 2015 Jan 27. PMID: 25620373.

Kevill JL, de Souza FS, Sharples C, Oliver R, Schroeder DC, Martin SJ. DWV-A Lethal to Honey Bees (Apis mellifera): A Colony Level Survey of DWV Variants (A, B, and C) in England, Wales, and 32 States across the US. Viruses. 2019 May 9;11(5):426. doi: 10.3390/v11050426. PMID: 31075870; PMCID: PMC6563202.

Le Conte Y, de Vaublanc, Crauser G, et al. Honey bee colonies that have survived Varroa destructor . Apidologie 38, 566–572 (2007). https://doi.org/10.1051/apido:2007040

Locke, B. Natural Varroa mite-surviving Apis mellifera honeybee populations. Apidologie 47, 467–482 (2016).

Locke B, Thaduri S, Stephan, J G et al. Adapted tolerance to virus infections in four geographically distinct Varroa destructor-resistant honeybee populations. Sci Rep 11, 12359 (2021).

Luis, A.R., Grindrod, I., Webb, G. et al. Recapping and mite removal behaviour in Cuba: home to the world's largest population of Varroa-resistant European honeybees. Sci Rep 12, 15597 (2022). https://doi.org/10.1038/s41598-022-19871-5

Martin, S.J. (1995) Reproduction of Varroa jacobsoni in cells of Apis mellifera containing one or more mother mites and the distribution of the cells. J. Apic. Res. 34(4), 187-196

References and further reading

Martin S J, Kemp D, (1997) Average number of reproductive cycles performed by Varroa jacobsoni in honey bee (Apis mellifera) colonies, Journal of Apicultural Research, 36:3-4, 113-123, DOI: 10.1080/00218839.1997.11100937

Martin S J, A population model for the ectoparasitic mite Varroa jacobsoni in honey bee (Apis mellifera) colonies, Ecological Modelling, Volume 109, Issue 3, 1998, Pages 267-281, ISSN 0304-3800, https://doi.org/10.1016/S0304-3800(98)00059-3.

Martin, S J, Varroa destructor reproduction during the winter in Apis mellifera colonies in UK. Exp Appl Acarol 25, 321–325 (2001).

Martin S J, Brettell L E, Annual Review of Virology, Vol.6:49-69, (2019), https://doi.org/10.1146/annurev-virology-092818-015700

Martin and Grindrod: Natural Varroa-Resistant Honey Bees: BBKA News Special Issue Series: 2020.

Martin, S.J., Hawkins, G.P., Brettell, L.E. et al. Varroa destructor reproduction and cell re-capping in mite-resistant Apis mellifera populations. Apidologie 51, 369–381 (2020). https://doi.org/10.1007/s13592-019-00721-9

Matthijs S, De Waele V, Vandenberge V, Verhoeven B, Evers J, Brunain M, Saegerman C, De Winter PJJ, Roels S, de Graaf DC, et al. Nationwide Screening for Bee Viruses and Parasites in Belgian Honey Bees. Viruses. 2020; 12(8):890. https://doi.org/10.3390/v12080890

McAfee A, (2021), The Legacy of the Mite-Biting Bees, American Bee Journal, https://americanbeejournal.com/the-legacy-of-the-mite-biting-bees/

Mitchell D, (2023), Honeybee cluster—not insulation but stressful heat sink, J. R. Soc. Interface.202023048820230488http://doi.org/10.1098/rsif.2023.0488

Möckel, N, Gisder, S., Genersch, E., Horizontal transmission of deformed wing virus: pathological consequences in adult bees (Apis mellifera) depend on the transmission https://doi.org/10.1099/vir.0.025940-0 (2011)

Mondet, F. et al. Antennae hold a key to Varroa-sensitive hygiene behaviour in honey bees. Sci. Rep. 5, 10454; doi: 10.1038/srep10454 (2015).

Mondet, F., Blanchard, S., Barthes, N. et al. Chemical detection triggers honey bee defense against a destructive parasitic threat. Nat Chem Biol 17, 524–530 (2021).

Mullin CA, Frazier M, Frazier JL, Ashcraft S, Simonds

R, Vanengelsdorp D, Pettis JS. High levels of miticides and agrochemicals in North American apiaries: implications for honey bee health. PLoS One. (2010)

National Bee Unit (2017), Managing *Varroa* leaflet.

Oddie, M., Büchler, R., Dahle, B. et al. Rapid parallel evolution overcomes global honey bee parasite. Sci Rep 8, 7704 (2018).

Omar R E M, Effect of Varroa Infestation on the Development of Body Weight and some Reproductive Organs of Honeybee Drones, Apis mellifera L. (2017)

Page R E Jnr, (1981), Protandrous Reproduction in Honey Bees, Environmental Entomology; 359-362, as cited by Thomas D Seeley in The Lives of Bees.

Payne A N, Walsh E M, Rangel J. Initial Exposure of Wax Foundation to Agrochemicals Causes Negligible Effects on the Growth and Winter Survival of Incipient Honey Bee (Apis mellifera) Colonies. Insects. 2019 Jan 8;10(1):19. doi: 10.3390/insects10010019. PMID: 30626042; PMCID: PMC6359559.

Ramsey SD, Ochoa R, Bauchan G, Gulbronson C, Mowery JD, Cohen A, Lim D, Joklik J, Cicero JM, Ellis JD, Hawthorne D, vanEngelsdorp D. Varroa destructor feeds primarily on honey bee fat body tissue and not hemolymph. Proc Natl Acad Sci U S A. 2019 Jan 29;116(5):1792-1801. doi: 10.1073/pnas.1818371116. Epub 2019 Jan 15. PMID: 30647116; PMCID: PMC6358713.

Rehm, S., and Ritter W.. . 1989. Sequence of the sexes in the offspring of Varroa jacobsoni and the resulting consequences for the calculation of the developmental period. Apidologie. 20: 339–343.

Remolina SC, Hughes KA. Evolution and mechanisms of long life and high fertility in queen honey bees. Age (Dordr). 2008 Sep;30(2-3):177-85. doi: 10.1007/s11357-008-9061-4. Epub 2008 Jun 22. PMID: 19424867; PMCID: PMC2527632.

Rinderer T E, De Guzman L I, Lancaster V A, Delatte G T, Seltzer J A, (1999) Varroa in the Mating Yard: I. The Effects of Varroa jacobsoni and Apistan® on Drone Honey Bees. Am. Bee J. 134–139

Rosenkranz P, Aumeier P, Ziegelmann B. Biology and control of Varroa destructor. J Invertebr Pathol. 2010 Jan;103 Suppl 1:S96-119. doi: 10.1016/j.jip.2009.07.016. Epub 2009 Nov 11. PMID: 19909970.

Rothenbuhler W C, Behaviour genetics of nest cleaning in honey bees. I. Responses of four inbred lines to disease-killed brood, Animal Behaviour, Volume 12, Issue 4, 1964, Pages 578-583, ISSN 0003-3472, https://doi.org/10.1016/0003-3472(64)90082-X.

Schiffer T, Natural Bee Husbandry, Issue 12 (2019).

Simone-Finstrom M, Spivak M, Propolis and bee health: the natural history and significance of resin use by honey bees. Apidologie 41, 295–311 (2010). https://doi.org/10.1051/apido/2010016

Spivak, M., Danka, R.G. Perspectives on hygienic behavior in Apis mellifera and other

References and further reading

social insects. Apidologie 52, 1–16 (2021). https://doi.org/10.1007/s13592-020-00784-z

Spivak M, Masterman R, Ross R, Mesce KA. Hygienic behavior in the honey bee (Apis mellifera L.) and the modulatory role of octopamine. J Neurobiol. 2003 Jun;55(3):341-54. doi: 10.1002/neu.10219. PMID: 12717703.

Spivak, M., Danka, R.G. Perspectives on hygienic behavior in Apis mellifera and other social insects. Apidologie 52, 1–16 (2021). https:// doi.org/10.1007/s13592-020-00784-z

Stevens C, VSH bee breeding, (https://www.stevensbeeco.com/).

Swindon Honey bee Conservation Group, founded by the late Ron Hoskins (www.swindonhoneybeeconservation.org.uk). This website is now defunct with some pictures available on Facebook. Content added from Honnigman P, (2015), Oxford Natural Beekeeping Group.

Topolska G. Varroa destructor (Anderson i Trueman, 2000); zmiana w klasyfikacji w obrebie rodzaju Varroa (Oudemans, 1904) [Varroa destructor (Anderson and Trueman, 2000); the change in classification within the genus Varroa (Oudemans, 1904)]. Wiad Parazytol. 2001;47(1):151-5. Polish. PMID: 16888966.

University of Sussex, Breeding hygienic honey bees, and the effect of hygiene on bee pests and diseases, https://www.sussex.ac.uk/lasi/sussexplan/hygienicbees#:~:text=During%20the%20past%20three%20years,queen%2rearing%20and%20breeding%20 activities.

Valentine A, Martin SJ (2023) A survey of UK beekeeper's Varroa treatment habits. PLoS ONE 18(2): e0281130. https://doi.org/10.1371/journal.pone.0281130

Villegas AJ, Villa JD, Uncapping of pupal cells by European bees in the United States as responses to Varroa destructor and Galleria mellonella (2006), Journal of Apicultural Research 45(3): 203–206

Visick O,. Ratnieks F., : Ancient, veteran and other listed trees as nest sites for wild-living honey bee, Apis mellifera, colonies, October 2023: Journal of Insect Conservation https://doi.org/10.1007/s10841-023-00530-7

White J, (2019), BBKA Magazine, June edition, pages 201-202.

Further reading

Seeley, TD, The Lives of Bees, (2019), Princeton University Press.

Heath D, Treatment-Free Beekeeping, (2021), IBRA and Northern Bee Books.

Education and science-based website: www.varroaresistant.uk

Westerham Beekeepers website: https://westerham.kbka.org.uk/

Appendix

Short-term biotechnical stepping-stone

From engaging with beekeepers locally and at presentations around the country, it has become obvious that there is great interest in transitioning away from miticide treatments. This interest, however, is matched by a deeply engrained fear that "*if you don't treat, your bees will die*". When we started the project at Westerham Beekeepers, we were no different. Colony survival was more important than colony fitness.

This appendix is included as an optional stepping-stone towards Varroa-resistance for beekeepers. The approach provides a period of time to refocus on the interaction between the honey bee and Varroa, rather than which miticide treatment to use at different points of the season.

With thanks and credit to Dr. Ralph Büchler for these ideas who inspired and mentored us through the first phase of our project. We first met Ralph at Gormanston (summer bee school just north of Dublin) in 2017. He was giving a series of lectures on Varroa spanning 4 days, where we were introduced to 2 techniques that controlled Varroa levels on the winter bees. Queen frame trapping and summer brood removal. Details of both can be found at the Westerham Beekeeper website (https://westerham.kbka.org.uk/natural-beekeeping/), including how to make a queen frame trap for a fraction of the commercial cost.

Biotechnical versus miticide treatment

The research that convinced us to try these methods was coordinated by the Kirchhain Institute in Germany across 140 colonies. The test was to see how the colonies fared under different late summer Varroa regimes, by comparing the size of the colonies in the following spring, as a proportion of the size in the previous summer, before Varroa reduction had taken place – a measure of how they had overwintered.

There were 3 Varroa reduction methods tested:-

1. A control, which used slow released formic acid, a popular method in Germany.

2. Brood removal in August, where frames of sealed brood were replaced with drawn comb for the queen to lay winter bees. Frames of stores were retained for the colony. This takes under 10 minutes per colony.

3. Queen frame trapping, where the queen is restricted to laying up brood comb in a cage, which attracts the colony's Varroa. Once sealed, the brood frame is discarded with the mites and replaced with another for the queen to lay up and the process repeated. There were 2 versions tested. One started in July and the other in August.

The results (see Fig.17.1) showed considerably larger colonies in the spring using queen frame traps, with positive implications for the coming season's honey yield and bee production. Interestingly, the formic acid treated colonies and those where all of the sealed brood was removed in August came out about the same the following spring.

Appendix

Fig. 17.1: Comparison of overwintering colonies after different mite reduction strategies the previous summer.

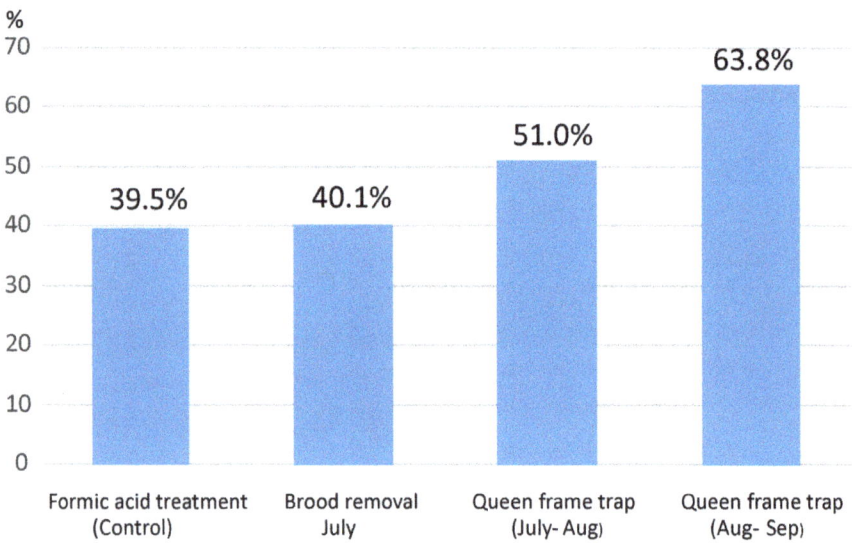

Chart: Research from the LLH Bee Institute, Kirchhain.

Westerham Beekeepers' experience

We trialled these methods in 2018 for half of our colonies and continued with a few colonies into 2019. Here are some summary thoughts.

Queen frame trap ("QFT")

The QFT ran from around 10th July for about 4 weeks, with the queen caged on two consecutive brood frames over that time. The queen was then released back into the colony to lay the winter bees in August, with very low levels of Varroa.

Fig 17.2: The Westerham queen frame trap built for a fraction of the commercial cost and very efficient at removing Varroa.

Pictures: Steve Riley.

Brood removal

This started around 10th-15th August. It took under 10 minutes per colony, but forethought was required to source clean drawn comb for the queen to start laying winter bees immediately. All frames of stores remained.

We found this approach sufficient for overwintering, but less effective at mite reduction than the queen frame trap, as Varroa on the bees were not caught.

Results

We were excited at the end of a nervous first winter to see our bees bursting out of the crown boards in spring. 26 out of 28 colonies had got through. The 2 colony losses were due to autumn queen issues, after the inspections had ceased. No losses from Varroa.

Appendix

We had started gathering mite drop data to try and understand the seasonal trends and impact of these techniques. The graph below illustrates the low level of mites in the 2018/19 winter resulting from the July-August 2018 QFT and brood removal exercises. In the summer of 2019, the effectiveness of the mite reduction techniques can be seen in the colonies.

Fig 17.3: Illustrates the effectiveness of two biotechnical methods of removing Varroa – more effective than miticide treatments. No winter treatments were applied.

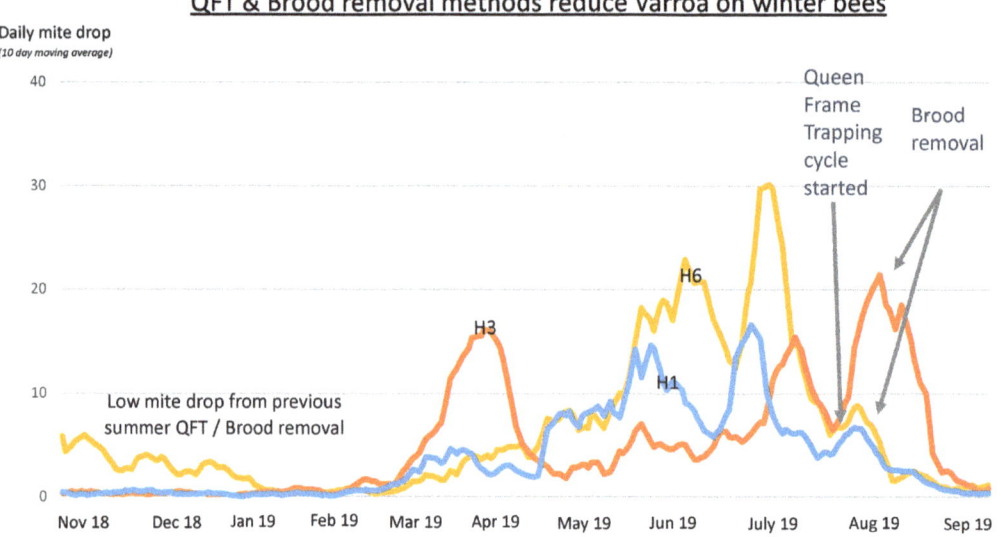

Graph: Steve Riley.

This is one of two Westerham Beekeeper slides that Ralph used in his series of four presentations to the National Honey Show given in October 2019. The lectures are free to view at https://www.youtube.com/watch?v=KwuR3uMkMF0

The lead group only used this Varroa reduction strategy for a season or two on some of the colonies before phasing it out. We could see how it worked and could always resort to it if all else failed. We were also uncomfortable with killing brood, although appreciated that it would be heavily parasitised by Varroa. Fundamentally, the methods were so successful at removing Varroa, more so than any combination of miticides we had previously used, that in effect, they had replaced treatments and taken the mite problem away from the bees to handle. They weren't the long-term solution but had shown us that miticides were not required in beekeeping.

The biotechnical strategy also served the purpose of overcoming the fear of stopping miticide treatments and to move on to finding resistance traits in our colonies. What was useful in that transition period were the eight months or so after the Varroa reduction had taken place through to the following April, where the honey bees and Varroa were co-habiting without a December treatment. The strong brood development in the spring provided an abundance of opportunities for Varroa to reproduce and for honey bees to stop them. That was our cue to closely monitor mite levels and for signs of uncapping and chewed-out pupae to interrupt Varroa reproduction.

We were then in a position to draw conclusions on the ability of different colonies to handle Varroa, with a focus on colony fitness, not just survival. The short-term transitional stepping stone had worked.

www.ingramcontent.com/pod-product-compliance
Lightning Source LLC
Chambersburg PA
CBHW041244240426
43670CB00025B/2983